The Rise of Rhythm Studies

The Rise of Rhythm Studies

Mediating Dimension, Discipline, and Scale

Edited by
Mark Lussier and Richard C. Sha

BLOOMSBURY ACADEMIC
NEW YORK • LONDON • OXFORD • NEW DELHI • SYDNEY

BLOOMSBURY ACADEMIC
Bloomsbury Publishing Inc, 1359 Broadway, New York, NY 10018, USA
Bloomsbury Publishing Plc, 50 Bedford Square, London, WC1B 3DP, UK
Bloomsbury Publishing Ireland, 29 Earlsfort Terrace, Dublin 2, D02 AY28, Ireland

BLOOMSBURY, BLOOMSBURY ACADEMIC and the Diana logo are trademarks of Bloomsbury Publishing Plc

First published in the United States of America 2026

Cover design by Eleanor Rose
Cover image © Flavio Coelho / Getty Images

A catalogue record for this book is available from the British Library.

A catalog record for this book is available from the Library of Congress.

ISBN: HB: 979-8-7651-2528-1
ePDF: 979-8-7651-2531-1
eBook: 979-8-7651-2530-4

Typeset by Integra Software Services Pvt. Ltd.

For product safety related questions contact productsafety@bloomsbury.com.

To find out more about our authors and books visit www.bloomsbury.com and sign up for our newsletters.

For Jonathan Culler

Contents

Figures

Acknowledgments

Mark Lussier would like to thank his tireless co-editor, Richard Sha, who long ago endorsed the concept of two conference sessions focused on rhythm (a prospect that ended with the coming of Covid-19). Richard thereafter embraced the idea of a collection, sustaining our efforts, and during our many exchanges across time and space he taught me more than books ever could. Mark also thanks Marcia (the rock of ages in life and love) for her patience and support. Richard Sha thanks Mark for coming up with the idea for the volume in the first place, for his boundless energy sustaining its production over the past three years, and for teaching him about Buddhist practices along with many other rhythmic phenomena. He also thanks Arlene (*Cuore mio*), without whom nothing matters. More than a thousand emails later, and weekly meetings over Zoom and in person, here it is. At Bloomsbury, we thank Amy Martin for her enthusiasm for the volume and her patience during its long gestation; Hali Han for trying to keep us on track; and Sara Brunton for her meticulous copy editing. Mark and Richard thank our wonderful contributors whose labors of ages herein provide abundant recompense, by providing food for future years.

Introduction—The Rise of Rhythm Studies: Mediating Dimension, Discipline, and Scale

Mark Lussier and Richard C. Sha

Rhythm, in its diverse forms, provides the threads that connect and condition the interplay of interiority and exteriority, no surprise given that "everyday life remains shot through and traversed by great cosmic and vital rhythms: day and night, the months and the seasons, and still more precisely biological rhythms" (Lefebvre 82) that traverse this seeming divide. Such an active state of rhythmic manifestations and operations within and without creates opportunities "for thinking about contemporary social and [cultural] phenomena in terms of rhythm" (Crespi and Manghani 11) along numerous plateaus. Rhythm thus understood can serve simultaneously as both "the organization of sense in discourse" (Meschonnic 82) and the organization of energy across space (Nevejan and Sefkarti 233–5) and time. Indeed, rhythm must be understood as already existing in sense and space (a proposition at the foundation of Henri Lefevre's work), as its physical, physiological, and phenomenological attributes continually bind and cross these discrete zones of becoming, when we speak "at length of different kinds or types of rhythm, and of all rhythms (in plural) that surrounds us" (Barletta xxvi). Past becoming distends the present into the future.

The essays that follow certainly attest to the emergence of rhythm as animating and connective principles at work in our bodies and throughout the surrounding cosmos. However, from its inception and development and through subsequent applications and refinements, rhythm came to be seen equally (even necessarily) as an instrument of investigation, while remaining a manifestation of force within varied forms and phenomena. Rhythm as an efficacious method within a wide range of disciplines led to cross-disciplinary observations of complex operations. The disciplines represented by the authors herein—aesthetic and artistic analyses, affective, critical and literary studies, sociological data and extrapolation, and science studies—evoke a widened scale of rhythm across our disciplines, a very wide something observed and observed something. This observation speaks to the ways that those in disciplines spanning art history, literature, neuroscience, affect theory, and sociology grapple with differing modes and understandings of mediation, of discipline, of dimension, and of scale. As we are increasingly aware, rhythm is a vehicle to break out of patterns and structures (both within and without/inside and outside) and across scales.

The very omnipresence of rhythm, its dynamics and dynamism, its diverse forms, its connections to forms of energies and engagements within the self and society (within culture writ large) facilitate critiques of traditional agency by transforming these energies into *eperon* or spurs (Derrida). The *eperon* or spur, like the prow of a ship, cleaves a path through water, thus serving as a tool to enhance movement through analytic, cognitive, linguistic, stylistic, or other material impediments. Rhythm often operates as an agent capable of cutting through ossified disciplinary operations, exchanging past for current (or future) rhythms, and encouraging the circulation of time with space. Cultures across time are replete with such paths of distributed agency emerging from the embrace of the rhythmic as both explanatory vehicle and sharp edge. Two historical examples (one inside and one outside of the body but both within emergent scientific investigation) help establish rhythm as primal/primary agent of change that can gesture toward its possibilities as a concept and mechanism and style for large-scale transformation and understanding. At its best, rhythm provokes nothing less than recut epistemes, leaving us wondering why the original cuts were there to begin with.

The discovery of rhythm inside the body confirmed that interiority was not separate from exteriority; rather, inside and outside share dynamic principles that shape one another in what we now describe as rhythms. Early in the seventeenth century, alchemist, philosopher, physician, and scientist Robert Fludd, for example, offered a complex vision of the cosmos founded exclusively on rhythmic operations (*c.*1609) and provided detailed designs like "Universal Monochord" to underscore commonality rather than difference along plateaus of investigation and discovery. Shortly thereafter, one of Fludd's colleagues and dear friends, William Harvey (*c.*1628), established definitively the circulatory flow and rhythmic patterns of blood through the science of hydrostatics and experimental studies of animals. If fluid flow could help make sense of the body and its circulation, then inside could no longer be othered from the outside. Harvey's observations established the hydraulic one-way flow of blood, yet his

De Motu Cordis encountered considerable opposition because he established rhythm as a binding principle. No longer a mystery, the truths of inside are revealed by medical physics, and such physics makes the inside part of the material world. Given that medical physics broadly served as a key part of standard medical training, Harvey's findings were eventually confirmed by subsequent experiments conducted by towering figures in physiology, including Robert Boyle (of Boyle's law), Robert Hooke (of Hooke's law), the philosophers Thomas Hobbes and John Locke, Benjamin Franklin, and Antoine Lavoisier.

Almost two centuries later, Thomas Young's experimental demonstration of the wave aspect of light demonstrated once again that the rhythmic operations of the external world mirrored those of the internal. Moreover, in demonstrating that matter functions as a wave, Young made it possible to see the world as being comprised of rhythm instead of hard individuated corpuscles. In fact, his famous double-slit experiment, resulting in the discovery of patterns of interference in its spectra that could only be explained by light as wave, unwove a hardened and inflexible Newtonian particulate view of physical dynamics and established experimentally a rhythmic element capable of functioning as the foundation of material operations within and without. This discovery liberated scientific thinking from reductive Newtonian dynamics, although it drew the ire of future prime minister Henry Brougham, who excoriated Young for indulging in hypotheses without evidence in the pages of the *Edinburgh Review*, which he founded. Young's insights, however, proved definitive and drove new patterns of rhythmic thinking, leading to the eventual unification of electricity and magnetism by James Clerk Maxwell at mid-century and therein established oscillatory and vibrational forces at work within the expression of physical dynamic phenomena. Under the aegis of rhythm, types of physical phenomena like electricity and magnetism were no longer discrete but interconnected.

Both Harvey's work on circulation and Young's waves of light help us envision rhythm as a transformational vehicle with affective and effective implications. Both inaugurated paradigm shifts that continue to inform current understanding of exteriority and interiority of both our bodies and the surrounding environment in which we are embedded. Harvey's observations set the stage for modern medicine, and as one philosopher of science suggests, "Modern medicine would be impossible without Harvey's discovery" (Sethna). The quandary of light as wave versus particle provided the physical context for the emergence of relativistic effects and quantum mechanics in the opening decades of the twentieth century (Gamow). Simply put, the act of measuring changes what is being measured, making an objectivity separate from subjectivity exact a price. As Kuhn argued, some anomalies grow up into a new paradigm, and we are witnessing rhythm becoming paradigmatic.

What changes as a result of a rhythmic approach to phenomena? As a way to foreground the interplay of inner and outer as a potentially transformative experience of rhythm, let's now consider a concrete and typical practice at the core of daily life for a Buddhist. Most practitioners of meditative techniques, whether experienced monks with extensive training or novitiates first coming to contemplative practices, learn and come to know that the method at the foundation of stability resides in the exchanging of self for other, specifically the practice of *tonglen*. The training begins with a fundamental use of the technique of creative visualization; every inward breath is accompanied by an imaginative process of ingesting the pain and suffering of both people and the world. Thereafter, every outward breath visualizes relief from the pain and suffering that pervades the world, giving this back to exteriority. Ideally, this meditative process undercuts patterns of selfishness and cultivates compassion within the self and the world even as each breath immerses self into the world.

This exchange, this receiving and sending at the core of meditative practices, finds its stability in an understanding that exteriority and interiority condition each other, with the body and mind of the practitioner serving as the vehicle within which the inner and outer interpenetrate through rhythmic processes. The technique at the heart of the method depends upon a process of internal rhythm, respiration, to implicate inner and outer through the individual practitioner, thereby promoting transformation within and without, thereby modeling a means to overcome the tendency to focus on the self, versus all that encloses self-consciousness, and provides a mechanism into/through the wider world. This singular and foundational rhythmic process achieves its sought ends within the context of other forms of rhythm: the circadian process of sleep and awareness, the periodicity of daily practice, the social commitments of our work, and the fundamental in-and-out breathing at its core. Such forms of rhythm often expand beyond this discrete practice (meditation), as one exchanges the solitary practice for movement into and through the wider world. Indeed, the goal might be to extend this rhythmic practice and its outcomes (individual stability) into other rhythmic forms during daily interactions within a workspace (through mindful management and compassionate engagement with colleagues). Indeed, the outcome of such daily practices allows the opportunity to be awake at work and to seek in such work the relief of pain in suffering of one's colleagues and our students.

Seen in this light, rhythm has transformative power to redefine agency as an interaction and immersion with our surround. At its best, agency, the ability to act, is a chance to engage in environmental play. Minimally, through rhythm, agency becomes a set of interwoven practices, with rhythmic coordination allowing individual practice to meld with collective ones. As we see agency as an interaction, we empower ourselves to move away from desired outcomes that we can learn to sense are not in sync with the world. We can thereby learn to see societal effects instead of myopic individual agency. Affectively, a rhythmic approach to life enables play with the world as opposed to against it. In this way, inside and outside are in constant interplay, and the cut between them becomes the invitation to egoism.

Rhythm has the potential to redress the balance between the power of awareness, which can be overstated, and the agential aspects of the nonconscious, which we tend to undercount. Katherine Hayles defines the nonconscious as "a broad environmental scanning that operates below conscious attention" (10). Attention to rhythm suggests patterns nested within other patterns, but that nesting need not tell the story of domination. The facet of conditions that we tend to forget is that they are not nouns but rather verbs. Conditioning in this view is an interaction, not a superimposition. If rhythm helps establish the boundaries among phenomena and also the kinds of interactions that are possible, then it shapes agency. However, that our internal rhythms interact with worldly rhythms does not in any way imply predetermination; rather, we must reconsider how agency works in relation to external rhythms. Rhythms remind us to temper our sense of individualism as we are, like it or not, part of a larger whole.

Recent work on consciousness has documented that what was before considered awareness is more and more the work of pre-personal mechanisms. Anyone who has driven somewhere without realizing how they got there recognizes the degree to which nonconscious cognition supports consciousness. Or consider the tennis player who is asked to explain his forehand but can't because it has become part of muscle memory and routine. What rhythm offers is a means of thinking about how awareness and the nonconscious interact, and that agency requires a combination of both. The fact that we may have no actual awareness of how our consciousness works does not prohibit rhythms from coordinating. Within neuroscience, the turn to thinking about the coordination of specific frequencies of gamma and theta rhythms, for instance, shifts our attention from individual neurons to their coordinated rhythms and interacting populations. Modularity and localization are being replaced by coordinating rhythms and interactions between brain regions; in this way we are empowered to think about how functions get spread through the brain.

Changing Scholarly Habits

Carolyn Pedwell redefines the scale of meaningful change by considering the revolutionary power of the rhythm of daily habits. From David Bissell she redeploys habits as "competencies that temporality possesses us" (xvii), warning that habits exceed human calculation and control. And as the process of habitual conditioning is underwritten by rhythm, should one consider rhythm as an objective phenomena or rather as the medium through which the cuts of significance are made? Pedwell's reminder that temporality possesses us serves as a caution that we must work with it and within it. Vincent Barletta's response is to ask the degree to which our activities shape space and time. Rhythm, he will not allow us to forget, takes hold of us. It is the wave of the mind anterior to experience that conditions experience itself.

Henri Lefebvre argues that rhythm requires a transdisciplinary approach (94). Under rhythm, nature is not only no longer carved at the joints, but the location of any joints is unclear. According to Lefebvre, a rhythmanalyst studying the Mediterranean town is "strictly speaking neither psychologist, nor sociologist, nor anthropologist, nor economist; however, he borders on each of these fields in turn" (94). We should note that in this view, rhythm becomes the discipline, the key factor enabling the analysis to hang together because no one discipline corners the market on all the tools needed for the shed. Nor should any one sense dominate: in a rhythmic view, music is not just a sonic experience, but rather invites the entire body (Hamilton, Macarthur, Squires, Rugby, Wiseman, 17).

For Pascal Michon, rhythm departs from metrics, and in a return to the ancient Greek *rhythmos*, rhythm becomes about time's very flow. Instead of breaking time up into strict and preconceived units of strong and weak beats, a strictly metrical approach, an attention to flow focuses on how individuation itself occurs, leading to the very production of significance (23–4). In this way significance and meaning are not already present; they require rhythm to come into being. In short, meaning is mediated by rhythm.

Rhythm thus demands more attention to flow and less worry about being captive to it, as it is ideally an interaction. At its best, flow offers a seamless blend of inside and outside. In his famous study of creativity, Miyhali Csikszentmihalyi documented that creative individuals of all kinds share the experience of flow, a state of consciousness when things appear to be going automatically but, in actuality, are the product of intense attention and concentration (110). As a result, there is often "immediate feedback to one's actions" (111), a sense of balancing between external challenges and skills (111), and action merges into awareness, allowing distractions to recede (111–12). Self-consciousness seems to disappear (112), as self and surround and activity become one.

Sociologists have embraced rhythm for its ability to make present the routines of the everyday, and for its ability to coordinate time with space. Literary scholar Jonathan Culler considers how rhythm enlists us in process itself (22), suggesting that rhythm is about the spread of agency. Through time-lapse photography, Dawn Lyon captures the daily rhythms of the Billingsgate Fish Market, and in the process, gives insight into the making of sociality itself. Literary scholar and cultural historian Saidiya Hartman turns to "the rhythms of everyday life" (*Wayward Lives* 4) in slums to imagine how to bring to life impoverished neighborhoods and marginalized lives for whom there are only scattered diaries and photographs. Those still fragmented shots are coaxed into a rhythm that allows crowds of different races to converge onto and within urban spaces. In this way, urban slums breed lives. Tracey Potts reimagines urban spaces as spaces for the organization around "work and leisure" (36), a reimagining that requires an attention to rhythm. For Caroline Levine, "rhythm refuses distinctions between aesthetic form and other forms of lived experience" (53), a refusal that allows us to attend to the temporal patterns within institutions and what she calls "the political life of form" (80).

If rhythm allows spaces to be woven into time and back again, context is transformed from a backdrop into a way of thinking about how routines actually facilitate the cuts of space and time. Vincent Barletta examines routines that bring into being space and time itself. Yi Chen describes new tools like polyrhythmia, which open up context to multiplicity and dynamism, and make rhythm not about domination, but rather about parallel lives and polysemy. For Henri Meschonnic, rhythm is about the very organization of significance itself, and in this way, sound can no longer be split from sense, and rhythm itself is transformed into a means of sense-making. We have only begun to outline the transformative potential of a rhythmic approach to phenomena, ranging from consciousness through physical interactions to urban spaces.

Although we find especially promising the power of rhythm to provoke more genuine interdisciplinary pluralisms, we would be remiss if we did not attempt to squarely face rhythm's challenges. How can rhythm possibly live up to its hype? As Kuhn recognized, one paradigm change only anticipates the next one. Haun Saussy quotes Lisa Gitelman: "the success of all media depends at some level on inattention or 'blindness' to the media technologies themselves" (87), which is to say that when rhythm is the medium, success and blindness are bound together.

The problem is exacerbated by the very proliferation of versions of everyday rhythms: put simply, rhythm appears everywhere. Percy Shelley might worry that "we have eaten more than we can digest." More detail should perhaps not be confused with different and more capacious details. When rhythm becomes about sense-making and the very generation of significance, how to reign in hubris? Although we admire Saidiya Hartman, the role of rhythm in what she calls "critical fabulation" is tricky. One the one hand, rhythm helps constitute the wayward and enslaved lives lost to historical archives. On the other hand, how to remain critical in light of rhythm's constitutive power to make the marginalized present in the rhythms of space? How to leverage rhythm as opposed to being seduced by it, especially when our awareness of it is often quite limited? How to prevent the flow/jouissance from rendering us blind? Recognizing the task as impossible because fact bleeds into fiction, which in turn bleeds into fantasy and violence, Hartman nonetheless remains committed to writing the narratives of lives otherwise irretrievably lost, while at the same time, "jeopardizing the status of the event" ("Venus" 11). Yet to what extent is it the work of rhythm itself to do precisely that? When the cure becomes automatic, we need to be wary.

Contents

We offer this collection of essays as a similar spur: we hope to advance rhythm studies by opening up new sites of intervention and new ways of coming to terms with rhythm's mediating powers. Our insistence on rhythm's mediation is designed to curb excess optimism about what rhythm can do. Essays by Caroline Levine, Chiara Bassetti, Keith Leonard, Nick Admussen, and Richard Sha highlight rhythm as process. For Caroline Levine, the discourse of creativity has ignored the rhythms of routine, in spite of the fact that creativity would languish without it and the forms of routine lent themselves to radical experimentation. She argues that even when these routines are imposed from without, they are critically enabling for marginalized groups. Chiara Bassetti demonstrates the intersubjective work of rhythm within dance and argues that rhythm can mediate hierarchy into relationality. Keith Leonard powerfully shows us how Black poets such as Evie Shockley and Nathaniel Mackey turn to rhythm to counter Blackness as social death by generating alternative forms of time outside Western clock time. Nick Admussen explores an influential underground Chinese journal, *Jintian*, within its polyrhythms of production, reception, and imitation. When rhythms become the fulcrum through which we read, texts proliferate what he calls "shapes in time" allowing for multiple spaces for the polis to reside. Richard C. Sha turns to Frantz Fanon and the mechanisms of brain waves to reexamine the relation of rhythm to epistemology. For him, rhythm is about the gap between feltness and knowing, a plenitude but uncertain epistemological space Hartman calls "Black noise" ("Venus" 12).

Chapters by Sha Xin Wei, Anna Gibbs, and Nina Kraus consider rhythm for its potential to help us think in terms of feedback loops. The exploratory chapter by Sha Xin Wei takes an experimental approach to rhythm, one that demonstrates the costs of phenomenological accounts of it that would ironically hermetically seal it into a completed process, making Merleau-Ponty's inside yet another version of the objective body. The journey is exhilarating, as the otherwise homogenous experience of time and space is transformed into a lived encounter with the potentialities of difference, what Wordsworth might have referred to as "spots of time." With Anna Gibbs, we undergo a different kind of exploration: one that enables us to rethink capitalism's destructive legacies by considering how it literally bends time out of joint, making time itself dehumanizing. Finally, if only all neuroscientists could write as engagingly and knowledgably as Nina Kraus. An expert on sound and its manifold and surprising influences on the brain, she demonstrates the rich cognitive work that rhythm enables but is nonetheless taken for granted.

Still others frame rhythm as new forms of presencing and being: consider the chapters by Georges Roques, Yolaine Escande, and Mark Lussier. Georges Roque takes seriously how and why painters such as Kandinsky, Klee, and Mondrian turned to rhythm to rethink the work of painting. In a larger view, he asks why there was so much collective interest in rhythm between 1910 and 1930. Yolaine Escande unpacks the rhythmic work of the brushstroke in Chinese landscape painting, and in the process helps us to feel in palpable ways what artists such as Lu Ji, Hwang Gongwang, Leng Qian, and Wang Meng are doing. Last but hardly least, Mark Lussier considers how rhythm transforms boundaries into generative conditions; in his wide-ranging chapter from physics through medicine to cosmology, he reminds us of the conceptual advances made possible by a rhythmic approach.

Works Cited

Barletta, Vincent. *Rhythm: Form and Dispossession*. U of Chicago P, 2020.

Chen, Yi. *Practicing Rhythmanalysis: Theories and Methodologies*. Rowman and Littlefield, 2016.

Crespi, Paola, and Sunil Manghani, eds. *Rhythm and Critique*. Edinburgh UP, 2020.

Csikszentmihalyi, Mihaly. *Creativity: Flow and the Psychology of Discovery and Invention*. HarperCollins, 1996.

Culler, Jonathan. "Why Rhythm?" *Critical Rhythm: The Poetics of a Literary Life Form*. Edited by Ben Glaser and Jonathan Culler. Fordham UP, 2019, pp. 21–39.

Derrida, Jacques. *Spurs: Nietzsche's Styles/Eperons*. Translated by Barbara Harlow. U of Chicago P, 1981.

Gamow, George. *Physics: Foundations and Frontiers*. Prentice Hall, 1960.

Hamilton, Andy, et al. "Dialogue on Rhythm: Entrainment and Dynamic Thesis." *The Philosophy of Rhythm: Aesthetics, Music, Poetics*. Edited by Peter Cheyne, Andy Hamilton, and Max Paddison. Oxford UP, 2019, pp. 15–42.

Hartman, Saidiya. *Wayward Lives, Beautiful Experiments*. W.W. Norton, 2019.

Hartman, Saidiya. "Venus in Two Acts." *Small Axe: A Caribbean Journal of Criticism*, vol. 12, no. 2, 2008, pp. 1–14.

Hayles, N. Katherine. *Unthought: The Power of the Cognitive Nonconscious*. Chicago: U of Chicago P, 2017.

Kuhn, Thomas. *The Structure of Scientific Revolutions*. U of Chicago P, 1962.

Lefebrve, Henri. *Rhythmanalysis: Space, Time and Everyday Life*. Bloomsbury, 2004.

Levine, Caroline. *Forms: Whole, Rhythm, Hierarchy, Network*. Princeton UP, 2015.

Lyon, Dawn. "Doing Audio-Visual Montage to Explore Time and Space: The Everyday Rhythms of Billingsgate Fish Market." *Sociological Research Online*, vol. 21, no. 3, 2016, https://doi.org/10.5153/sro.3994

Meschonnic, Henri. *Critique du Rythme: Anthropologie historique du langage*. Verdier, 1982.

Michon, Pascal. *Rythmes, pouvoir, mondialisation*. Presses Universitaires de France, 2005.

Nevejan, Caroline, and Pinar Sefkatli. "City Rhythms: An Approach to Urban Rhythm Analysis." *Rhythm and Critique*. Edited by Paula Crespi and Sunil Manghani. Edinburgh UP, 2020, pp. 218–30.

Pedwell, Carolyn. *Revolutionary Routines: The Habits of Social Transformation*. McGill-Queens UP, 2021.

Potts, Tracey. "Life Hacking and Everyday Rhythm." *Geographies of Rhythm: Nature, Place, Mobilities and Bodies*. Edited by Tim Edensor. Routledge, 2016, pp. 33–44.

Saussy, Haun. *The Ethnography of Rhythm*. Fordham UP, 2016.

Sethna, Duhn. *The Wine-Dark Sea Within: The Turbulent History of Blood*. Basic Books, 2022.

Wordsworth, William. *The Prelude. Book XII*. Edited by Jonathan Wordsworth, M.H. Abrams, and Stephen Gill. W.W. Norton, 1979, pp. 428–29.

Part One

Theoretical Foundations

1

Between Feltness and Knowing: Fanon, Brain Waves, and the Epistemology of Rhythm

Richard C. Sha

On the one hand, rhythm can deliver feltness before knowing.[1] Rapid increases in pulse and in rates of breathing are certainly felt, but feeling is not necessarily knowing. Does one attribute these changes in physiological rhythm to the traffic or to the mortgage that needs to be paid? Feltness can certainly feel like knowing. The current turn to emotion and affect within so many fields is predicated on them being a kind of cognition, with cognition understood as a kind of automatic information processing, which allegedly confers on the species survivability. To challenge this premise, consider how well automatic processing deals with climate change or political polarization. Cognition has considerably narrowed the gap between emotion and knowledge, but it does so at the expense of mistaking information for knowledge. Information is raw data, but knowledge is about how to apply the data. Feeling fear does not mean that the reasons that we attribute to the fear or what we think is the source of the fear are true. Our current emotional turn accepts automaticity as the necessary price for efficiency, but this comes at the costs of knowledge, deliberation, and the ability to reevaluate.[2]

On the other hand, although rhythm can be a powerful means to interaction and coordination, rhythm is not necessarily felt with awareness. Rhythm can

1 Ben Glaser thus opens *Critical Rhythm* with the warning that to reflect in poetic rhythm is "not to seamlessly or properly engage the native or collective rhythms of a linguistic and cultural heritage. It is to encounter rhythm (through meter) as a defamiliarized and defamiliarizing force" (2). He goes on to urge that rhythm cannot always sustain the "boundary game" it plays between the literary and the aesthetic (2).
Thanks to Marshall Alcorn for organizing the study group of the Forum of Psychiatry and the Humanities around Fanon, and to Jean Wyatt for leading the discussion on Fanon that sparked this paper.

2 See Ruth Leys, *The Ascent of Affect*.

be seductive without our knowing it (Culler, *Lyric* 134). There are also many kinds of rhythm that we cannot feel directly, and even when we do feel rhythms directly that does not imply that the meaning of those rhythms is stable. We have no direct access to our brain rhythms, for instance. Although we can know about our brain rhythms, what does the fact that they cannot be felt mean, especially as Jonathan Culler warns that although rhythm "appears to be a property of systems, yet it is above all an (impersonal) experience" ("Why Rhythm?" 28, 34)? My wager here is that from the standpoint of epistemology, rhythm is about the gap between feltness and knowing and the limitations of human experience.[3] When rhythm encourages social interaction, if this is without awareness and at the subpersonal level, how much credit to grant rhythm for social change because it is partly automatic? Framing the problem in this way raises the issue of the value of awareness; the kind of knowing that comes with rhythm; and the role of dual process theories in epistemology. Briefly, dual process models separate unconscious and automatic actions from conscious and deliberate ones, arguing not only that most of consciousness relies on automatic affective processes but that they are processed separately from slow, deliberate, and effortful processes.

Increasing evidence, however, shows that there are overlaps between dual systems, and that the relation between automatic and conscious is not one of kind, but one of degree.[4] The overlap matters because if the automatic and conscious are mixed, then the one no longer precludes the other. Conscious attention is not necessarily antithetical to skilled action; for concert pianists and athletes to improve performance, focused attention to and awareness of their own performances is highly necessary.[5] To the extent that our thinking is hybrid, both automatic and intentional, intentionality has a wider orbit than has been recognized because automaticity does not preclude intention. Add to this, as we shall see with Frantz Fanon, the fact that our sense of the meaning of rhythm can change. If intentionality coopts automatic processes, then criticism will have to be more thoughtful about the meaning of rhythm as a foundational process. Consciousness can no longer be cast aside as evolutionary detritus.

Within studies that understand emotions to be appraisals, the idea that our emotions are shaped by how we assess the situation allows cognition and

[3] For a model study of rhythm that avoids "the naiveté of direct reporting" and "foregrounds a constitutive distance between orality and its observers," see Haun Saussy (5–15).

[4] On dual process theories, see Chaiken and Trope, Hayles, and Leys (*Anatomy of a Train Wreck*).

[5] See Leys, *Anatomy* 155–58.

emotion to become intertwined so much so that one cannot be extricated from the other. Appraisal theorists of emotion insist that when I fear, I fear something, and therefore my fear is a quick, often automatic, appraisal of the situation. Where appraisal theorists of emotion connect emotions to goals, goals are often reduced to representations that work automatically.[6] J.A. Bargh insists that goal representations work without any conscious will; constructs can be directly activated by the environment. Two of his most famous examples are: feeling warmth when holding a hot beverage primes one to feel favorably toward another person (Bargh 65), and words connoting old age unconsciously influence research subjects to walk more slowly. Ruth Leys warns that social psychologists have recently made goals a kind of automatic information processing without intention (Leys, *Anatomy* 29), and even worse, they adhere to a cognitive science that understands automatic and intentional actions to be mutually exclusive.[7] As Leys demonstrates, social psychologists have been unable to replicate these famous priming experiments (*Anatomy* 206–16), casting doubt on the basic premises of automatic goal pursuit and on the ability of automaticity to do away with conscious intent.[8] For our purposes here, all this reliance upon automatic information processing without intention further lowers the stakes of what counts as knowledge. In many cases, understanding is out of the picture. Again, having information does not tell me what to do with the information. Moreover, phenomenal immediacy is blind to all the construction that makes such immediacy possible. Although automatic processing may not feel effortful, there is significant effort involved.[9] The implications for the collapse of dual process

[6] Automaticity is far from a simple concept. Some automatic processes require an attentional allocation. Others rely on a purported automatic perception–behavior coupling, a kind of monkey see, monkey do. Moreover, consciousness itself may be predictive and constructive. On the problems associated with automaticity and the emotions, see Uleman and Bargh 5–15. Early in his career John Bargh acknowledged that some automatic processes "require conscious awareness and attentional processing" (Uleman and Bargh 10). More recently, he understands conscious processing to be a liability. "Our everyday experiences active ideas and desires and even ways of thinking about the world" (Bargh 79). Leys (*The Ascent*) credits appraisal theorists with aligning emotion with intentionality, but if the goal is automatic, then appraisal theorists too can be downplaying the conscious aspects of emotion. As she recognizes if mimicry is automatic, why is it selective – why do we not mimic everything we see? (*Anatomy* 182.)

[7] See for instance Gollwitzer and Bayer, who argue that our minds can be divided into "deliberative mindsets" and "implemental mindsets." Are these mutually exclusive?

[8] See also Schimmack's *Replicability Index*.

[9] The reason why the effortful and effortless distinction matters is that the unconscious system is supposed to save effort and feel effortless. Just because something feels effortless does not mean that there is an absence of effort. Take for instance the immersion in a task described as flow. See Justin London on the problems of perception endemic to rhythm. Our apprehension of rhythms is limited to those between the 100 milliseconds to 2 second range (174), and our "perceptual system structures the auditory stimulus" (179).

theories for rhythm studies suggest that we must proceed with caution before the scalar leaps from metrics to culture and from information to knowledge.

The epistemological problems surrounding rhythm only get worse. Karl Figlio argues that certainty itself is a feeling "through which we believe and through which our experience becomes evidence" (12). He continues "certainty is neither an empirical finding nor even a logical conclusion, but a state of mind—a state of mind that does not tolerate the complexity of either empirical investigation or of conceptual analysis" (12). When certainty is moved away from reason and toward emotion, the feeling of certainty threatens knowing rather than consolidates it, and consequently the path from feeling to knowing seems more like an abyss. According to Arie Kruglanski, the conviction of certainty shuts down the process of inquiry and removes the possibility of alternative hypotheses (21–2). With regard to rhythm, when we collapse feltness with knowing, we run the risk of crediting rhythm with a kind of knowledge it may not have and shutting off the search for knowledge prematurely. Even worse, because emotion today conditions cognition, what to do when our emotions get it wrong? This problem worsens with today's rampant hatred, polarization, and anxiety. Figlio warns that certainty encourages psychological projection, the confusion of my intentions with yours, facilitating, for instance, projection of hatred onto the hated object. Our emotions are out of whack. Something is rotten, and not just in Denmark. Add to this the way in which polarization encourages certainty, because the choice is not a real choice. Consider how the Democrat will not accept the Republican's views as a choice, and the possibility opens up that polarization, not to mention rhythm, does our thinking for us.

The feltness of rhythm thus offers fewer truths that can be taken to the bank than we might think.[10] Even worse, our awareness of rhythms is limited; we can and often are seduced by it; and, when perceived, it is subjectively perceived. My modifier "subjectively" is not a damnation, but rather a kind of filtering to reflect upon.[11] What we take to be the meaning of rhythm, moreover, can change. The slippage between emotion and knowledge adds a further monkey wrench. For

[10] Phillipe Lacoue-Labarthe warns against the mythic or "legendary" use of rhythm, citing Hölderin's "All is rhythm [Rhythmus]; the entire destiny of man is one celestial rhythm" (cited in Lacoue-Labarthe 139–40).

[11] See Cheshire Calhoun's feminist defense of subjectivity as a kind of knowing. She questions the blanket use of subjectivity to empty out knowing and proffers the idea of "biographical meaningfulness" to capture kinds of subjective knowing like personal expertise.

instance, when we are hangry, we often do not realize that what we are perceiving as anger is really low blood sugar. One should never appeal one's parking ticket at lunch time, because judges might mistake their hunger for legitimate reasons to be angry. In her increasingly influential constructionist model of emotions, Lisa Feldman Barrett argues that our emotions are in part about our body budgets. As part of our iteroceptive system, emotions function in part to make us aware of our homeostatic states. In this view, "a bad feeling does not mean something is wrong. It just means you are taxing your body budget" (Barrett 76). Yet because this model suggests our emotions straddle our bodily needs and the external world, it is difficult to disentangle whether our emotions are speaking to our inner states or to our relations with the world. Feltness is thus wrapped up in the possibility of error. With rhythm, this problem is compounded. Our inability to parse the difference between internal and external rhythm offers significant challenges to what we should do in light of them. Making matters worse, our "body-budgeting regions are sluggish to correct their predictions" (Barrett 77).

I now turn to two case studies of rhythm to explore how rhythm raises problems for epistemology and for the relationship of feeling to knowing.[12] Rhythm is exciting precisely because it suggests a feltness that entangles space with time. But what is the meaning of this feltness, and how can we know that meaning? What to do with the gap between feltness and meaning, which entanglement does not solve? (Entanglement, as Thomas Young found, is very tricky to know. It still rubs against much of our experience of the world.) Moreover, rhythm studies tend to emphasize rhythm's possibility over constraints. My first case study therefore considers, on the one hand, Frantz Fanon's depiction of colonialism as the disruption of any kind of innate bodily rhythm, and on the other hand, his turn to the rhythms of Senegal's Léopold Senghor and his fellow Martinican's Aimé Césaire to learn how to ascend from those depths. What does it mean to be deprived of bodily rhythm? Homi Bhabha writes in his foreward to *The Wretched of the Earth* that Fanon's "great contribution to our understanding of ethical judgment" is his attention to the "psycho-affective realm" (xix).[13]

[12] David Nowell Smith models rhythm as posing itself as a problem, and notes that it becomes "critical" with Kant's insistence upon the subjective conditions of experience itself. "Rhythm … becomes critical in the wake of philosophy reformulating itself as critique" (42). The difficulty is that "critical" here refers both the limits of experience but also the transcendence of those limits. Hence, he urges that we think of rhythm not as critical but as speculative (42).

[13] Fanon's designation of a "psycho-affective realm" unites affect theory with the psyche and psychoanalysis.

When colonialism works at this level, it can disconnect one from one's own body and bodily rhythms. And then why does the poetic rhythm of negritude inspire Fanon, and how does that rhythm begin to overwrite the loss of bodily integration with the world?

Rhythm is also exciting because it is a powerful engine of collectivity. The phenomenon of entrainment suggests that our sensory motor system resonates with certain rhythms, confusing perception and action. Shared rhythms across bodies intimate that rhythm can be a way of social organization. My second case study concerns brain rhythm. Because we know that rhythm plays a key role in brain function but have no direct access to the feel of these rhythms, feltness in this case runs the danger of becoming an epiphenomenon. The challenges that brain rhythm pose to epistemology is the inaccessibility of their feltness; and with this loss, the loss of direct means of their coordination, not to mention the end of that coordination. What does this unfelt coordination mean, and can this unfelt coordination have an ethics? What does a social collectivity without intention mean? When groups respond to a collective rhythm, does that hold any ethical weight, especially if this response is automatic? Moreover, automaticity exacerbates the problem of seduction. Brain rhythms furthermore impose constraints on cognition, and it is important to stress both the possibility and the pitfalls.

Taken together, then, my two case studies suggest that rhythm is best seen as an ideal whose claims to knowledge need interrogating; especially so when rhythm underwrites an ethical claim because potentialities for both good and ill must be accounted for, along with opportunities and constraints. Because we experience value through emotion, this leap from feeling to ethics is especially dangerous. Only by putting a pause on the leap from rhythm's feltness to rhythmic knowing can rhythm fully realize its potential to reorganize disciplinary fields, what they know, and sociality itself. After all, if we are compelled by rhythm into sociality, then we can be pushed into both belonging and tribalism, not to mention fascism. Entailed in this pause is a reevaluation of automaticity, an end to the presumption that the automatic and conscious are separate systems, and a knee-jerk casting aside of intentionality because it seems irrelevant. For Fanon, there is almost nothing automatic about rhythm. He must weigh whether it hurts or helps. As much as he might want the idea of a pure Black rhythm that can restore a kind of bodily integrity that colonialism has taken away, it is a myth whose purchase comes at a steep price.

Fanon

In his influential 1952 study *Black Skin, White Masks*, Franz Fanon writes "in the white world (*dans le monde blanc*) the man of color (*l'homme de couleur*) encounters difficulties (*rencontre des difficultés*) in the development of his bodily schema (*dans l'élaboration de son schema corporel*)" (*Black Skin, White Masks* 110; *Peau Noire, Masques Blanc*, 126). Remarkable here is Fanon's suggestion that colonialism is so disruptive that it disconnects the colonial subject from his or her very body. In this view, the disconnection is so severe that rhythm is unavailable without distortion. His logic goes like this: Our interaction with the world is built upon proprioception, the perception or awareness of the movement of the body, and bereft of a bodily schema, this fundamental orienting power to the world is missing, making it nigh on impossible to integrate oneself with the world. When the "development of his body schema" stops, "the man of color" is no longer capable of being in synch with the world because that schema is part of a process that has been disrupted. He is "enfermé" (*Peau Noire, Masques Blanc* 125) or enclosed as it were outside of his own body. "Assailed at various points, the corporeal schema crumbled (*s'écroula*), its place taken by a racial epidermal schema (*cédant la place à une schema corporel*)," Fanon warns (*Black Skin, White Masks* 112; *Peau Noire, Masques Blanc* 127). Surface replaces depth, and body becomes schematized into race: both flattened and unmoored. The results are devastating.

The chapter in which these piquant remarks appear has been badly translated into English as "The Fact of Blackness." Fanon's French original is "Expérience Vécue du Noir." The difference matters because Fanon underscores lived experience over facticity. "Facticity" misnames the problem; Blackness is not as simple as what is a fact, but rather, it is about how to have experience in a world that would claim as fact that Black people are incapable of having experience at all (Gordon 48–9). Moreover, in lived experience, the line between feeling and knowing is both fraught and ambiguous because of a temporal gap between the felt and conscious understanding of the felt. Simply put, racism and colonialism are felt on the pulses, but what it means shifts in Fanon from an unmooring, through a felt inferiority, to a white projection. There is a further problem. When Fanon adds that "consciousness of the body is solely a negating activity" (*Black Skin, White Masks* 110), awareness is not associated with a higher consciousness, but rather with the cultural assimilation of one's inferiority that one is only aware of to a degree. In this view, lived experience cannot be reduced to fact because in the living of the experience, one has to sort out what this experience means. A major problem with dealing with awareness is that it does not always help. In a colonized or racist world, awareness disables. So the larger question that Fanon raises is what can we do with awareness, and how can we make it work for the oppressed as opposed to against them?

Fanon eloquently captures this fundamental disorientation by figuring himself as a mollusk. He depicts himself making "progress by crawling" (*Black Skin, White Masks* 116). If we recall his self-description as "enfermé," he is self-enclosed as in a shell. Later in the chapter on "Expérience Vécue du Noir," Fanon describes his own body as having a pseudopod. Fanon has metamorphized himself into a mollusk whose false foot extends itself outside the shell from time to time. Pseudopodia function for locomotion and anchoring. Yet rather than helping him to explore the world as it does for the mollusk, his false foot "secreted a race" (*Black Skin, White Masks* 122). Fanon's remarkable full description is worth recalling:

> in the face of this affective ankylosis of the white man, it is understandable that I could have made up my mind to utter my Negro cry. Little by little, putting out pseudopodia here and there, I secreted a race. And that race staggered under the burden of a basic element. What was it? Rhythm! (*Black Skin, White Masks* 122)

> devant cette ankylose affective du Blanc, j'aie pu décider de pousser mon cri nègre. Petit à petit et lànçant çà et là des pseudopodes, je sécrétai une race. Et cette race tituba sous le poids d'un élément fondamental. Quel est-il? Le rythme. (*Peau Noire, Masques Blancs* 134)

Rhythm is Fanon's punchline, perhaps because as Jonathan Culler notes, that when we find it "rhythm enlists us in process in ways that other texts do not" ("Why Rhythm," 22).[14] Fanon's figure for rhythm is fantastic, highlighting the disfiguring work of colonialism and of racism, and forcing his readers to scratch their heads. Here, "rhythm" is proximate to a burden. On the one hand, rhythm is associated with Fanon's possible "negro cry," and thus might seem to be a positive and an incipient form of protest. On the other hand, when Fanon uses "j'aie pu," he turns to the past subjunctive, which indicates something hoped for or wished for, although in the past. The cry is no longer wished for. The mere fact of the cry being labeled "negro" is already an internalization of the racism that conditions the legitimacy of colonialism and thus reinforces the past tense

[14] Culler cautions that interpretative efforts surrounding rhythm are perhaps best seen as attempts to justify rhythm's hold over us (26).

subjunctive. The French "nègre" oscillates between Black and the racial epithet that cannot be named, and the pastness of the wish suggests that the cry is no longer considered appropriate. That Fanon figures a race as a secretion of a false foot, renders it an excretion rather than a bodily essence, corporeal waste. The secreting is in the past preterite form, also highlighting the full termination of the act in the past at a specific point in time, but one nonetheless capable of haunting the self in the present. And while the chapter title insists on the color noir, which is an improvement over nègre, Fanon's cry is explicitly racialized into *nègre*. The idea of a racist call and response furthers the impact of rhythm, but negatively. Even *Noir* is only a category under the gaze of whiteness: "Car le Noir n'a plus à être noir, mais à l'être en face du Blanc" (*Peau Noire, Masques Blancs* 125). The Black no longer has to be a Black being, implying he or she has only Black being in the face of a white person. Fanon's "car" or "because," moreover, makes clear that he or she is beholden to the gaze of whiteness for being itself, and this alone is reason to be wary. Which is to say that whiteness refuses Blackness as genuine being. In this view, a racist Blackness cancels out being because without the white gaze, it cannot sustain itself. Blackness must refuse non-being.

To get to the core of rhythm, we need to look even more closely at Fanon and his contexts. Rhythm is embedded in cultural matrices that make it very difficult to parse what it stands for, surrounding it with ambiguity. Add to that Fanon's own declared ambivalence about the idea of any pure rhythm. The danger of any pure Black rhythm is a nostalgic primitivism. How can a Black man under the power of the white gaze or "le regard du Blanc" consider himself like a mollusk with a crooked pseudopod? Fanon begins by positioning himself "in the face" of the white man's affective ankylosis, but how does face meet foot? Does the pseudopod of a mollusk even have a joint that might stiffen? The foot is false, making it a kind of phantom limb, thereby intensifying its haunting power. In French, the false foot is thrown (*lancer*), suggesting a kind of arbitrary placement rather than a seamless move that helps one glide through space. Fanon staggers (*tituba*) on the burden (*sur les poids*), and the rhyme between the "tituba" and "poids" creates a drag, an echo in time, that doubles the feltness of the burden. The burden of an ankylosis or stiffening of the joints is an appropriate figure for a sedimentation into the past. Fanon's modifier "affective" suggests a conditioning power of emotion: the white man's ankylosis is contagious. Contagion means neither white nor Black is spared the disfigurement of colonialism. We therefore might note the affective traffic between white and Black, the subjective and the objective, which makes it difficult to track who is doing what to whom along with the limits of contagion. When rhythm is understood as a boundary condition that is contagious, therefore, its meaning and implications are far from clear.

Through affect, colonialism in this view, borrowing from Sara Ahmed, writes the borders between colonial subject and colonist (44–6). As such, anything like innate rhythm can only lead to confusion because the borders between bodies are being written through affect. As he descends down the great chain of being, Fanon, the mollusk, is close to entering the zone of non being, what Fanon refers to as "transparent, not there, disappeared" (*Black Skin, White Masks* 112) because one can do what one wants to a mollusk with few ethical qualms. As Lewis Gordon argues, in Fanon, there is self, other, and nonbeing. Where the "other" still has the potential to partake of the human, nonbeing is a total negation of the human. This twofold negation makes ethics irrelevant because one can do whatever one wants to a nonbeing. In a world where nonbeing is a category into which human beings can be thrust, race can and does function as a secretion of pseudo-being. The idea of ankylosis or stiffening of a false foot causes staggering, but the false foot only appears because racism makes it impossible for Black bodies to navigate the world because it denies Black bodies humanity and converts them into phobic objects. Fanon's outlandish metaphor, it turns out, could not be more apt. Its very excess mirrors racism's negation of both otherness and being itself: through nonbeing, Blackness is twice removed from humanity and fixed. Hence the need for a pseudopod to navigate this world, and hence the need for a shell to enclose oneself from the white gaze. Hence the false foot secretes a race, the false embodiment that allows Black nonbeing contact with the world. Fred Moten argues that for Fanon, the Black cannot be an other for another Black (178), making any contact with the world even more tenuous.[15]

If the violence of colonialism and white supremacy can only be figured in such a wildly distorted bodily map, and if that distortion is accompanied by a rhythm that is near to the burden of race, how might rhythm become the engine of possibility instead of something that disables? Staggering returns us to the loss of proprioception, but what does it mean for race to impact proprioception? And is race the cause of the staggering, or does the staggering happen because an

[15] Moten would insist on thingness as opposed to non being. He situates Fanon's idea of Blackness in the impossible social and fugitive space between the juridical/pathological and the social, between "the facts" of what's wrong with Blackness and the lived experience of Blackness (179). This betweenness makes Blackness for Moten's Fanon "irredeemably homeless" (187).

original rhythm is disturbed or obfuscated by race? How to promote staggering as disruption as opposed to disability? Where Senghor considered negritude to be a direct and lyrical experience of the world, Fanon usually positions negritude in the gaze of the Blanc, whiteness. If staggering is the result of the contest between an original rhythm and a racist rhythm, then the staggering reads less as a disability and more as an inhibition to the colonialist or racist project, a registering that under colonialism that the center cannot hold. And yet because subject and object are intertwined, the answer is both. Fanon's lyrical repetitive play between *petit* and *petit* and *lançant çà et là* hints at the inertial pull of a deformed language to entrap the racialized Black subject. *Çà et là* repeats but reverses the sounds in *lançant*, making language into a palindrome, a cul-de-sac. Once again the action of the rhythm traps into an endless looping rather than enables. It remains to be seen whether the loop can be consigned to the past, especially given Ian Hacking's concept of the "looping effect," the ability of clinical designations to make up and change people.[16]

We shall soon see why the action of the rhythm entraps. As Fanon will claim, Blacks are a "phobogenic object, a stimulus to anxiety" (*Black Skin, White Masks* 151). In French, he writes "Le nègre est un objet phobogène, anxiogène" (*Peau Noire, Masques Blancs* 156). When reduced to an object that produces an irrational fear and anxiety, the raced body becomes a fixed form of nonbeing and lacks agency to such a degree that it cannot even move its own body in space. Important to grapple with is the problem of what side rhythm is on. Is rhythm necessarily part of the problem or can it be part of the solution? The false foot declares rhythm as problem, one intensified by its falseness, its secretion of race, and its twice removal from humanity. Crucially, in this view, Blackness is without agency and, to boot, stuck with the emotional labor of grounding white anxiety.

Because Blackness is represented as the stimulus but without its own power, it is denied any agency. After all, what is a stimulus that lacks even the power even to be its own cause? This is of course how Fanon insinuates Blackness to be a white projection. If we recall Figlio on certainty and projection, the two consolidate each other. Fanon continues: "at the extreme, I should say that the Negro, because of his body, impedes the closing of the postural schema of the white man—at the point, naturally, at which the Black man makes his entry into

[16] Hacking insists "we do not have a choice not to use human kinds" (365) in part because they open up possibilities for human action (369).

the phenomenal world of the white man" (*À l'extrême, nous dirons que le nègre par son corps gêne la fermeture du schéma postural du Blanc, au moment naturellement où le Noir fait son apparition dans le monde phénoménal du Blanc*) (*Peau Noire, Masques Blancs* 162). Blackness here is said to be a blindspot in the phenomenal picture of the white man, but one which prevents phenomenological closure itself. This claimed blindspot enables Blackness to become phenomenologically relevant, but how does one become conscious of a blindspot because one sees through it? In this view, racism is so difficult to gain a purchase on because it sutures both the self and the phenomenological wholeness of reality. Fanon labels this blindspot as a psychosis: "whenever there is a psychotic belief, there is a reproduction of the self" (*Black Skin, White Masks* 161n25). Racism is psychosis, a simultaneous disconnection from and connection to reality, but one that makes reality itself into a hallucination. Thinking of racism as a psychosis allows one to explain why it is difficult to do without it and why one must do without it. The fissure in ontological wholeness provokes racism as a psychic and stronger reproduction of the self. Without this psychosis and hallucination of the other, amounting together as a blindspot, the self is unable to reproduce itself. Even worse, as Brian Massumi argues about fear, the mere fact of feeling fear becomes evidence that the basis of fear is justified, and this error is one reason why the gap between feeling and knowing must always be taken seriously even within the affects (56–9), and why rhythm cannot be presumed to dissolve epistemological problems.

At this juncture, it may be useful to return to the gap between feeling and knowing by thinking about the distance between information and what to do with the information, knowledge. Rhythm cannot mean one thing because it has been disturbed by colonialism; it sits at the intersection between the colonist and the colonial subject's body; it leads to a behavior like staggering, which can either be a positive or a negative; and it moves between fear and disgust ("la crainte et le dégoût," *Peau Noire, Masques Blanc* 158). Fanon's insight into phobogenesis makes race affectively define the borders between subjects. He comments: "in the phobic, affect has a priority that defies all rational thinking" (*Black Skin, White Masks* 155). He adds,

> The choice of the phobic object is therefore *overdetermined*. This object does not come at random out of the void of nothingness; in some situation it has previously evoked an affect in the patient. His phobia is the latent presence of this affect at the root of the world; there is an organization that has been given a

> form. For the object, naturally, need not be there, it is enough that somewhere it *exist*: It is a possibility. (*il y a organization, mise en forme. Car naturellement l'objet n'a pas besoin d'être là, il suffit qu'il soit: c'est un possible*). (*Black Skin, White Masks* 155; *Peau Noire, Masques Blanc* 158)

And yet how to sense the disturbance and to attribute meaning to the disturbance so one knows what to do in its wake, especially when the disturbance is also responsible for the awareness of rootedness? How to convert affective information into knowledge, when fear and disgust serve as a lightning rod for affect, organize it into form, and make objects out of mere possibility? When affects are considered automatic, one loses the ability to evaluate both the meaning of the fear that one feels and how it gets organized and even creates objects, and this loss leads to tragic consequences.

Senghor

Rhythm thus cannot be dislodged from the questions that surround it. At his nadir, Fanon considers the powers of negritude to restore his faith in a kind of intrinsic Black rhythm and to cancel out phobogenesis. He acknowledges the positive valence that negritude places on Blackness. He refers to "our singer, Léopold Senghor" (*Black Skin, White Masks* 122), the Senegalese President and ardent proponent of negritude and quotes his understanding of rhythm:

> It is the thing that is most perceptible and least material. It is the archetype of the vital element. It is the first condition and hallmark of Art, as breath is of life: breath, which accelerates or slows, which becomes even or agitated according to the tension in the individual, the degree and the nature of his emotion. This is rhythm in its primordial purity, this is rhythm in the masterpieces of Negro art, especially sculpture. (*Black Skin, White Masks* 122–23)

In the original French, Fanon cites Senghor:

> C'est la chose la plus sensible et la moins matérielle. C'est l'élément vital par excellence. Il est la condition première et le signe de l'Art, comme la respiration de la vie; le respiration qui se précipite ou ralentit, deviant régulière ou spasmodique suivant la tension de l'être, le degré et la qualité d'émotion. Tel est le rythme primitivement dans sa pureté, tel il est dans la chefs-d'oeuvre de l'Art nègre, particulièrment de la sculpture. (*Peau Noire, Masques Blanc* 134)

The pronoun "our" suggests at least a temporary affiliation between Fanon and Senghor. Senghor credits rhythm with feltness: sensibility, a tactical refusal of Blackness as an incapacity for experience. In a later work, Senghor argues that African Americans do not see objects; rather they feel them (*il le sent*) (259). This difference justifies Senghor's distinction between a European "raison discursive" and a Black "raison-toucher" (259–60). Feeling here is reracialized: what was an inability to feel becomes the cure to a dessicating rationalism. He then claims that rhythm is the primary condition and sign of art, likening it to respiration, which is in turn likened to an emotional quality. Rhythm builds a bridge between the object—*la chose*—and physiological vitality and emotion, but the bridges are only similies. Recognizing the insufficiency of figures, Senghor then grants rhythm a primitivity and purity, found in the masterpieces of Black art. And yet notice the tension between rhythm as a "condition première" and the dependent clause that begins with "suivant." Is rhythm primary or does it follow from the tension within the being and the being's quality of emotion? And if it follows, "suivant," how can it be either pure or primitive, as it is not its own origin? Even within Senghor's own promotion of the rhythms of negritude is a fundamental ambivalence that suggests consciousness of negritude is retroactive and a tactical move against racism. More importantly, if rhythms condition, it is itself also conditioned. Grounding art in physiology limits it to a bodily extension, but that extension is only necessary because the Black body itself is ungrounded by colonialism. Hence Fanon's term "secretion," a kind of corporeality, but one that is latent, and therefore initially ungrounding and later, one hopes, a form of waste.

Just how tight and lasting Fanon's affiliation with Senghor is, is unclear. Immediately after the quote, Fanon questions whether he has read the passage well: "avais-je bien lu?," he wonders. He adds: "je relus à coups redoublés" (*Peau Noire, Masques Blanc* 134): I reread again and again. What explains this insistence on rereading? Crucial here is the idea that rereading itself requires redoubled efforts. Fanon's choice of the simple past tense (*relus*) makes it an action completed once in the past, underscoring the intensified efforts it took to reread because there is no momentum to build upon.

Making rhythm even more difficult to parse, rhythm serves as the basis of Senghor's belief in negritude. In part, negritude exists to refute the tabula rasa theory of the Negro (*la théorie du table rase du Nègre*) that Senghor equates to "non-sens" (94), a negation of feeling itself. Even within his own life, negritude stands for a complex constellation of ideas. In his early career, Senghor valorized mixedness, and the idea that one could retain Blackness and assimilate. In the 1930s before political independence, cultural nationalism was necessary to revalue Blackness, and perhaps conditions political independence. Initially it frames Senghor's rejection of the superiority of French culture (Vaillant 155), but at the same time it has paradoxically an assimilationist bent. Ba argues that in the early days of negritude, Senghor insisted that "the Black man's task was to assimilate without letting himself be assimilated" (13), and this model perhaps allowed him to value hybridity and to see it not in terms of erasure.

In 1964 in *Négritude et Humanism*, Senghor took more care to distinguish between identification as "false assimilation" (44) and true assimilation, one which allowed for "emancipation intellectuelle et politique" (44). False assimilation was only identification with whiteness. In 1970, Senghor wrote to his biographer, Janet Vaillant, that negritude was not an essence but rather a "phenomenon and existence" (Diagne 123), but one could argue that it always hovered between the three. The stakes amount to the difference between ontology and epistemology: the former is about being, the latter includes the nature of experience along with the possibility of learning from experience. The epistemological frame around negritude allows it to name a being that functions differently in the face of different political challenges, and to avail itself of the learning that comes with experience. That Senghor's negritude itself hovers between Blackness and mixture further complicates the picture.

Senghor believes that in terms of poetry negritude refers to a unique rhythm on the part of Black poets, inflected by the rhythms of the tom-tom, the syncopation of the African drum. He writes, "le rythme, c'est l'architecture de l'être, le dynamism interne qui lui donne forme, le système d'ondes qu'il émet à l'adresse des autres" (281). Rhythm is the architecture of being, the internal dynamism that gives form, the system of waves that it emits to others. Vaillant argues that Senghor develops a dichotomy between white colonialists who practice "excessive ratiocination, egotistic individualism, and a grasping materialism" (156) and this was contrasted with Black "holistic comprehension, a sense of community, and a concern for a life of the spirit" (156). Yet Senghor does not think that negritude is something that only Black artists have: he credits Dvorak, Debussy, Picasso, Matisse, Victor Hugo, and Wolfgang Goethe with "des valeurs noirs" (Black values) (Celarent 303; Senghor 107). Hovering between degrees of difference and forms of assimilation, rhythm does not mark the boundary between Black and white; it negotiates that boundary, thus making it hard to determine whether it empowers or disempowers.

For all these reasons, Fanon actively questions whether feltness is knowing, and is engaged in a project of rereading and reevaluating the notion of Black rhythm. Praxis thus replaces doctrine. Fanon concludes: "Magic Noire, mentalité primitive, animism, érotisme animal, tout cela reflue vers moi. Tout cela caractérise des peoples n'ayant pas suivi l'evolution de l'humanité" (*Peau Noire, Masques Blanc* 137). "Black Magic, primitive mentality, animism, animal eroticism, it all floods over me. All of it is typical of peoples that have not kept pace with the evolution of the human race. Or, if one prefers, this is humanity at its lowest. Having reached this point, I was long reluctant to commit myself" (*Black Skin, White Masks* 126). Fanon thus foregrounds a reluctance to commit to rhythm: although it is an experience, it can and does mean many things; feltness requires episodes of reflection; changing political contexts means that rhythm needs to be adjusted, reunderstood, and redeployed. When is mixture an assimilation too far? One is reminded of Ralph Ellison's *Invisible Man* and the ten drops of black paint into a can of white paint, which only makes "the purest white that can be found" (202). That the paint company is "Liberty Paints" reminds us that national liberty too often comes at the expense of others. Whatever agency negritude affords Blacks, the fact that it reflects back over Fanon suggests that it makes passive, overwhelms, and thus must be guarded against. Any agency it gives is simultaneously taken away. Hence Fanon's insistence upon distance: it reflects back and thus is not the self. "Vers moi" (toward me) grants negritude an orientation toward, but one at best asymptotic to the self.

Césaire

Tactically, although negritude can help overcome a colonialist degradation of Blackness, it risks a naïve primitivism that eschews modernity. To wit, Fanon worries that taking Black rhythm at face value puts him in the camp of the irrational, and thus ultimately fulfills the colonist's stereotype of Blackness. As he moves from the negritude of Senghor to Césaire, Fanon writes: "je suis bâti d'irrationnel; je patauge dans l'irrationnel" (*Peau Noire, Masques Blanc* 135). "I am built of irrationality; I wade through irrationality." Now, can one be built of irrationality and wade through it simultaneously, which implies that one can move beyond it? The ontological fact of it is at odds with the acting in relation to it. Here again there is ontology and there is experience. Note that the one does not presume the other because experience shapes the meaning of one's felt Blackness. He continues: "Et maintenant, vibre ma voix!" (And now, my voice is vibrating!) What follows is a quotation from Césaire. Fanon's voice vibrates with Césaire's rhythms of negritude, an improvement over Blackness as pathology, but one not without dangers. Where Senghor had embraced mixture, Césaire's version of negritude treated mixture with suspicion, as an erasure of Blackness. At a conference for French writers, Fanon once signed himself in as Césaire. Fanon quotes Césaire, and his voice starts vibrating:

> Ceux qui n'ont inventé ni la poudre ni la boussole
> Ceux qui n'ont jamais su dompter la vapeur ni l'éléctricité
> Ceux qui n'ont exploré ni les mers ni le ciel
> mais ils savent en ses moindres recoin le pays de souffrance. (*Peau Noire, Masques Blanc* 135)

In these lines, Blackness is associated with the drumbeat of eight negations, and one positive: but they know every inch of the land of suffering. They neither invented gunpowder nor the compass. They never tamed steam or electricity. They never explored the sea or sky. Fanon concludes that this represents Blackness as humanity at its lowest. Césaire's drumbeat of "ceux," they, advertises the point of view of Blackness via distance. Fanon insists "all of it is typical of peoples that have not kept pace with the evolution of the human race" (126). Here the damning powers of primitivism could not be more clear. Fanon justifiably concludes, "toutefois, il fallait se méfier du rythme"; in every case, it was necessary to beware of rhythm, the friendship of mother earth, the mystical marriage of the group and the cosmos (*Peau Noire, Masques Blanc* 136). The price of this negritude and belonging is to be trapped, a cost apparently too high. Fanon insists: "we Negroes—backward, simple, free in our behavior. This is because for us the body is not something opposed to what you call the mind" (*Black Skin, White Masks* 126–27). Although this negritude is not walled off from mind, it is severed from modernity, and within its own walls, negritude has only bodily suffering as a resource.[17] It becomes the country (*le pays*) all to itself,

[17] On the problem of defining Africanness against modernity, see Olúfémi Táíwò. Basically, modernity cannot be just Western; to use decolonization to deny Africa modernity does not take African agency seriously; because colonization was an interaction between colonist and colonized, there can be no pure space for either.

a cost perhaps foregrounded in what Diagne calls Césaire's "Frenchest French" (127), a charge that reminds us that the "master's" tools have a price.[18]

And yet for both Senghor and Césaire, proponents of negritude as rhythm, the decision to situate négritude within the French language both highlighted the struggle from within which ran the risk of an all too successful assimilation. Here, perhaps to mitigate risk, Césaire's wicked pun on France within "le pays du souffrance" (suffering) (135), intimates that thus far the French legacy has been an asymmetrical suffering on the part of Black people, a souffrance that puts Black people *sous* France. The first part of the French word for suffering sounds like *sous*, or underneath. In this view, negritude risks perpetuating suffering and dependence, and the suffering will only end once Black people have gained independence. However, Fanon puts the subjective "one had to" in the past, opening the door to a future case without suspicion and putting an expiration date on his "toutefois," in every case. Moreover, Fanon highlights Césaire's play on "souffrance," which not only through the pun reasserts the mind's understanding of the body but also suggests that Black freedom demands a truly embodied mind. In this way the white conflation of Blackness with body is shown to be a loss.

Situated thus, Fanon's skeptical relation to negritude and to rhythm becomes the only justifiable position. As he will claim about Sartre: "the dialectic that brings necessity into the foundation of my freedom drives me out of myself" (*Black Skin, White Masks* 135). We return once more to the false foot driven outside the shell exploring the world but secreting race. Rereading in this view offers the only possibility that negritude in the future will enable more choices and agency, but rereading is far from offering a guarantee. At the same time, rereading is the precondition of negritude's ability to offer agency, because the meaning of being and of the body may have changed, and doctrine must give way to practice. Before independence, negritude as cultural nationalism may be the path to political independence. After independence, the amount of assimilation necessary to preserve autonomy may be more negotiable. Fanon recalls the wisdom of a friend in the United States who warns "the presence of

[18] Fanon returns to Césaire: Césaire learned first of all that "it is no use painting the foot of the tree white, the strength of the bark cries out from beneath the paint" (198). Here, Césaire's ability to see Blackness and whiteness as polarities even within a Black self grants negritude a tactical power that leads to the "murder" of the white parts of the Black self. Fanon writes, "once he laid bare the white man in himself, he killed him" (198). Yet Fanon argues "my unreason was countered with reason, my reason with real reason" (132).

the Negroes beside the whites is in a way an insurance policy on humanness. When the whites feel that they have become too mechanized, they turn to the men of color and ask them for a little human sustenance" (*Black Skin, White Masks* 129). Crucial here is that although Black rhythm functions to offer "human sustenance," its ability to insure humanness works primarily for whites since it cures them of an all too mechanistic reason. This is why Fanon admits that he "needed to lose himself completely in negritude" (135). And this is why Fanon cannot forget what Sartre did: "that the Negro suffers in his body quite differently from the white man" (138).

Brain Waves

We have seen Fanon raise basic questions about rhythm and the relation of feltness to knowing and about the epistemological rewards and downsides of awareness. For Fanon, it is often not clear whether rhythm is a burden or a resource. When it is a burden, neither is it clear how to move it into something enabling. Rhythm can be enabling and disabling simultaneously. And when it is a form of assimilation, how to know whether Blackness has been able to preserve any authenticity? I now want to shift gears to thinking about the potential of brain rhythms to alter our understanding of how we think. That we do not have any direct awareness of our brain rhythms does what to the relationship between feltness and knowing? If direct feltness is not possible, direct awareness is not possible. Without these, what is the ethical significance of rhythm?

Rhythms are exciting to brain science because they can help explain our cognitive limits, the complex coordinations of the brain, and how functional circuits might form (Watson and Buzsáki 68). In a larger view, within rhythm studies, more attention does need to be paid to the constraints imposed by rhythm. Earl Miller and Timothy Buschman argue that working memory, on the one hand, has a limited and discrete number of slots—no more than seven but an average of five ("Working Memory Capacity" 113–14). This explains why human beings on average can keep no more than five things in their heads at once. On the other hand, under a flexible resource model, information can be spread among objects, but with declining accuracy in proportion to the increase in objects. This model explains why even when subjects do not consciously perceive an object, they record a reduced amount of information about it (114). It also helps explain why human beings are so poor at multitasking. They proffer brain rhythms as a way to understand these inbuilt limits to awareness and to understand local to global communication in the brain (Watson and Buzsáki 68). Simply put, the brain uses oscillating waves to coordinate and spread its functions. Therefore, awareness itself may be the product of synchronized oscillations, an idea that strongly suggests brain function is tied to the coordination of brain rhythms. Rhythms within certain frequency bands might be more prone to coordinate. Because rhythms have an amplitude that take up a fixed amount of space, the idea that there are real limits to awareness becomes explainable. The geometry of brain rhythms may tell the story of our capacity limits: "only so much information can fit into an oscillatory cycle" (Miller and Buschman, "Brain Rhythms" 7). They continue: "any coding scheme based on repeated rhythm activity is naturally limited in bandwith" (ibid.). Miller and Buschman's insistence on rhythmic bandwidth reminds proponents of rhythm that they must address constraints as well as possibility.

All is not bad news, however. Synchronized brain rhythms increase the impact of neural spikes (Miller and Buschman, "Working Memory Capacity" 116) and have been correlated with attention and the functional work of sleep. "Increased attentional focus increases oscillatory synchrony between the visual cortical neurons" (116). Elsewhere, Miller and Buschman claim that synchrony improves the signal to noise ratio of neural signals ("Brain Rhythms" 2). Watson and Buzsáki connect oscillations to the capacity to learn new information so that it can shape future action (71). The fact that multiple brain rhythms are going on simultaneously can help us understand how the brain organizes many kinds of brain activity at once. Slower oscillations modulate the power of the faster ones (Watson and Buzsáki 69). Rhythm may also be the means to the broadcasting or spreading of local signals to the entire brain. That low-frequency brain waves coordinate with higher frequency ones suggests a movement away from single neurons and direct anatomical connections. Brain rhythms may be the key to how the brain scales up its functions. In fact, if rhythm helps the brain to form ensembles, "these ensembles can form, break up, and reform without changing the physical structure of the underlying network" (Miller and Buschman, "Brain Rhythms" 4), providing flexibility but not endless flexibility. In the face of all these brain rhythms, Donald Hebb's old neural saw "what fires together wires together" may need to be altered to say what oscillates together temporarily forms an alliance. If the old model relied on actual connections between neurons, the new model suggests momentary alliances built upon rhythmic coordination. Where neuroscientists in the past may have clung to models of modularity, new models might have to rely upon cooperation and coordination. Localization of brain functions may now need to give way to orchestration and rhythmic control. The upshot of neural synchrony might be cognitive flexibility, our ability to shift our thoughts and behavior (Miller and Buschman, "Working Memory Capacity" 116).

And yet what looks like cognitive flexibility or even creativity can seem from the other side a form of attention deficit disorder. Thomas Metzinger argues that for two-thirds of our waking lives, we lack mental autonomy, and he defines mental autonomy as the ability to avoid mind wandering. And yet if mind wandering is the means through which our brains work, it seems perverse to define mental autonomy in terms of the ways our brains cannot work. By contrast, Michael Corballis insists that mind wandering is the source of our creativity: it is "an extra resource to escape the here and now and the tasks at hand, and play" (147). Clearly the truth must lie somewhere between creativity and attention deficit. Once again, the turn to rhythm cannot wholly be a story of gain.

Assuming that we now can know that rhythms can enrich and perhaps explain our brain functions, and assuming that rhythm changes the models of how our brains work from localization and modularity of discrete signals to spreading and the involvement of populations of neurons, let us now consider the problem of our general lack of awareness of our brain rhythms. Simply put, we do not and cannot feel them. If rhythm owes its power to its experiential dimensions, then this is a very significant limitation.

And yet feeling is not the only form of knowing. Brain rhythms can be measured, tracked, and interrupted through transmagnetic cranial stimulation. Periodic electrical oscillations can be measured outside the brain (Watson and Buzsáki 68). Recent work on anesthesia has suggested that the loss of consciousness during anesthesia occurs because brain rhythms have become more rigid, and thus neurons are no longer excitable and thus are inert to stimuli (Kopell 1:01.53). The hypothesis is that anesthesia prevents the coordination of brain rhythms that lead to consciousness. Geometrical understanding of the waves associated with brain rhythms can provide fodder for arguments about cognitive capacity but also limits and disease. Kopell studies brain diseases such as Parkinsons, and she shows that electrical stimulation of brain rhythms can restore some of the body's mobility. Crucially, the turn to rhythm changes how we think about the brain and how we understand thinking itself. However, capacity is never without limits, a warning that becomes more salient in light of the tendency of neuroscience to insist that neural plasticity is about efficiency. The coordination of rhythms is temporary. Catherine Malabou reminds us that neural plasticity requires apoptosis or cell death and thus always comes at a significant price. Her larger point is that notions of efficiency are fantasies because they do not account for their costs (30–1).

Although it is easier to conceptualize neurons as individual actors, if rhythm is the vehicle of brain coordination, then the number of means to coordination increases well beyond the set anatomical pathways and single neurons. Yet these options too have their limits because of the amplitude of brain waves. In this view, rhythms may enhance the forms of plasticity and entrainment that are possible, but at the same time rhythms signal the limits of human thought.[19] All of this is to say that knowing the brain functions through rhythm does not yet tell us what these rhythms mean. Nor can it tell us the ethical or social implications of rhythm, even as it credits rhythmic ensembles as the engines of consciousness and of the social. We do not experience brain rhythms and even if we did, experience is not the same as comprehension (Culler, "Why Rhythm?" 32). For all these reasons, the epistemology of rhythm best remains an open question.

Works Cited

Ahmed, Sara. *The Cultural Politics of Emotion*. Routledge, 2015.

Ba, Sylvia Washington. *The Concept of Negritude in the Poetry of Léopold Sédar Senghor*. Princeton UP, 1973.

Bargh, John. *Before You Know It*. Touchstone, 2017.

Barrett, Lisa Feldman. *How Emotions Are Made: The Secret Life of the Brain*. Macmillan, 2017.

Calhoun, Cheshire. "Subjectivity and Emotion." *Thinking about Feeling: Contemporary Philosophers on Emotion*. Edited by Robert Solomon. Oxford UP, 2004, pp. 107–24.

Celarant, Barbara. "Review of Senghor, *Négritude et Humanisme*." *American Journal of Sociology*, vol. 19, no. 1, July 2013, pp. 299–305.

[19] Clayton thus argues that within musical rhythm "individual rhythms influence and entrain each other" (187), and that rhythms are "irreducibly social" (190). "The endogenous rhythms of which musical rhythm is built are biological processes" (196). My question would be what is the value of the social here, when rhythm is the unit of interaction or "interpersonal entrainment" (188), and what about the problem of seduction? Clayton provides two suggestive leads: metered music offers "future oriented attending," and the idea that "shared temporal flow is usually experienced as felicitous" (193, 194).

Chaiken, Shelly, and Yaacov Trope. *Dual Process Theories in Social Psychology*. Guilford P, 1999.

Clayton, Martin. "Entrainment and the Social Origin of Musical Rhythm." *The Philosophy of Rhythm: Aesthetics, Music, Poetics*. Edited by Peter Cheyne, Andy Hamilton, and Max Paddison. Oxford UP, 2019, pp. 183–98.

Corballis, Michael. *The Wandering Mind: What the Brain Does When You're Not Looking*. U of Chicago P, 2015.

Culler, Jonathan. *Theory of the Lyric*. Harvard UP, 2015.

Culler, Jonathan. "Why Rhythm?" *Critical Rhythm: The Poetics of a Literary Life Form*. Edited by Ben Glaser and Jonathan Culler. Fordham UP, 2019, pp. 21–39.

Diagne, Souleymane Bachir. "Rereading Aimé Césaire: Negritude as Creolization." *Small Axe*, vol. 48, November 2015, pp. 121–28.

Ellison, Ralph. *The Invisible Man*. Vintage Books, 1989.

Fanon, Frantz. *Black Skin, White Masks*. Translated by Charles Markmann. Grove P, 1967.

Fanon, Frantz. *Peau Noire, Masques Blanc*. Kiyikaat Editions, 2015.

Fanon, Frantz, *The Wretched of the Earth*. Foreward by Homi K. Bhabha. Translated by Richard Philcox. Grove P, 1963.

Figlio, Karl. "The Mentality of Conviction: Feeling Certain and the Search for Truth." *The Feeling of Certainty: Psychosocial Perspectives on Identity and Difference*. Edited by Nikolay Mintchev and R.D. Hinshelwood. Palgrave Macmillan, 2018, pp. 11–30.

Glaser, Ben. "Introduction," *Critical Rhythm*. Edited by Ben Glaser and Jonathan Culler. Fordham UP, 2019, pp. 1–17.

Gollwitzer, Peter, and Ute Bayer. "Deliberative versus Implemental Mindsets in the Control of Action." *Dual Process Theories in Social Psychology*. Edited by Shelly Chaiken and Yaacov Trope. Guilford P, 1999, pp. 403–22.

Gordon, Lewis. *What Fanon Said: A Philosophical Introduction to his Life and Thought*. Fordham UP, 2015.

Hacking, Ian. "The Looping Effect of Human Kinds." *Casual Cognition: A Multi-Disciplinary Approach*. Clarendon P, 1996, pp. 351–94.

Hayles, N. Katherine. *Unthought: The Power of the Cognitive Nonconscious*. U of Chicago P, 2017.

Kopell, Nancy. "Brain Rhythms in Health and Disease." CUNY Fifth Annual Sharon Cosloy and Edward Blank Family Distinguished Scientist Lecture, October 7, 2020.

Kruglanski, Arie. *The Psychology of Closed Mindedness*. Psychology P, 2004.

Lacoue-Labarthe, Phillipe. "The Echo of the Subject." *Typography: Mimesis, Philosophy, Politics*. Harvard UP, 1989, pp. 139–42.

Leys, Ruth. *Anatomy of a Train Wreck: The Rise and Fall of Priming Research*. U of Chicago P, 2024.

Leys, Ruth. *The Ascent of Affect: Genealogy and Critique*. U of Chicago P, 2017.

London, Justin. "Metric Entrainment and the Problem(s) of Perception." *The Philosophy of Rhythm: Aesthetics, Music, Poetics*. Edited by Peter Cheyne, Andy Hamilton, and Max Paddison. Oxford UP, 2019, pp. 171–82.

Malabou, Catherine. *What Should We Do with Our Brain?* Translated by Sebastian Rand. Fordham UP, 2008.

Massumi, Brian. "The Future Birth of the Affective Fact: The Political Ontology of Threat." *The Affect Theory Reader*. Edited by Melissa Gregg and Gregory Seigworth. Duke UP, 2010, pp. 52–70.

Metzinger, Thomas. "The Myth of Cognitive Agency: Subpersonal Thinking as a Cyclically Recurring Loss of Mental Autonomy." *Frontiers in Psychology*, vol. 4, December 2013, pp. 1–19.

Miller, Earl K., and Timothy Buschman. "Brain Rhythms for Cognition and Consciousness." *Neuroscience and the Human Person: New Perspectives on Human Activities*. Edited by S. Dehaene and W. Singer. The Pontifical Academy of Sciences, 2013, pp. 1–11.

Miller, Earl K., and Timothy Buschman. "Working Memory Capacity: Limits on the Bandwith of Cognition." *Daedalus*, winter 2015, pp. 112–22.

Moten, Fred. "The Case of Blackness." *Criticism*, vol. 50, no. 2, spring 2008, pp. 177–218.

Saussy, Haun. *The Ethnography of Rhythm: Orality and its Technologies*. Fordham UP, 2016.

Schimmack, Ulrich. *The Replicability Index*. https://replicationindex.com

Senghor, Léopold. *Négritude et Humanisme*. Éditions du Seuil, 1964.

Smith, David Nowell. "What is Called Rhythm?" *Critical Rhythm: The Poetics of a Literary Life Form*. Edited by Ben Glaser and Jonathan Culler. Fordham UP, 2019, pp. 40–59.

Táíwò, Olúfémi. *Against Decolonization: Taking African Agency Seriously*. Hurst and Company, 2022.

Uleman, James S., and John Bargh, eds. *Unintended Thought*. Guilford P, 1989.

Vaillant, Janet. "Review: Perspectives on Léopold Senghor and the Changing Face of Negritude." *ASA Review of Books*, vol. 2, 1976, pp. 154–62.

Watson, Brendon, and György Buzsáki. "Sleep, Memory & Brain Rhythms." *Daedalus*, winter 2015, pp. 67–82.

2

Rhythm: Inside and Outside the Head*

Nina Kraus

If sound loses time, it loses meaning.

My husband reads to me every night in bed before we go to sleep. With our special bear, Oatmeal, tucked in between us and listening too, this is a wonderful way to end the day and is a quotidian highlight. We deliberately choose books that are familiar—oft-read children's classics by E. B. White and the Harry Potter series make frequent appearances—so I do not worry about missing something important when I drift off. I have noticed that after some time—it can be as little as a few minutes if I am especially tired—the meanings of the words are gradually eclipsed by the sounds. I begin to hear sounds and rhythms instead of words and story, and the waxing and waning of the accents and stress patterns become a calming, lulling, treasured experience that soothes and resets me after a long day.

Why do we care about rhythm? It connects us to the world. It plays a role in listening, in language, in understanding speech in noisy places, in walking, and even in our feelings toward one another.

Rhythm is much more than a component of music. Nevertheless, music is probably what first comes to mind when we hear the word *rhythm*: drumming, jazz, rock and roll, marching bands, street performers with wooden spoons and five-gallon buckets, drum circles, time signatures, stomp-stomp-clap—we will, we will rock you—adventures on the dance floor, beatboxing, incantations, mantras, and prayers. Beyond music, we experience the rhythmic changes of the seasons. Some of us have menstrual cycles. We have circadian rhythms—daily cycles of mental and physical peaks and troughs. Frogs croak rhythmically to

* Updated and adapted by the author from chapter 6 of Kraus, N., *Of Sound Mind: How Our Brain Constructs a Meaningful Sonic World*. MIT P, 2021.

attract mates and change their rhythm to signal aggression. Tides, seventeen-year cicadas, lunar phases, perigees and apogees, are other naturally occurring rhythms. Human-made rhythms include the built world—street grids, traffic lights, crop fields, mowed designs in baseball diamond outfields, the backsplash behind the kitchen counter, spatial patterns in geometric visual artforms.

Maintaining rhythm is almost a biological imperative for some of us. My musician husband can get infuriated if we are playing together and I stop in the middle of a song. He needs to keep the pulse going. I have my own hiking version of this imperative. I need to *keep moving* no matter how tired I get, one foot in front of the other, even if I slow way down, or essential energy just drains away.

Music and rhythm are rooted in every known culture.[1] What parent does not use rhythmic rocking to soothe a crying baby? The repetitive sounds and silences that comprise rhythmic patterns make dancing possible, aid in the memory and reproduction of music, and facilitate group singing, playing, or drumming. Rhythm has been used for millennia to tie societal members together—the chants of a religious order or the cadence calls of military ranks are just two examples. Poetic works thousands of years ago, such as those of Homer, were chanted or sung, with rhythm serving a mnemonic function.[2] Repetitive or complex work engenders rhythmic accompaniment, in some cases to break the monotony, in others to actually help you perform the work better. Workers performing hard labor such as rock breaking chant to keep their sledgehammers swinging in rhythm.[3] Postal workers in Ghana hand-cancel stamps with a distinct rhythm.[4] Rug weavers in Iran use chants with a complex musical structure to communicate weaving patterns to their co-weavers.[5] All musical systems and styles have organizational rhythmic motifs. Indeed, the very universality of rhythm is a strong argument for the existence of biological processes governing the perception and production of rhythm.[6] Rhythms in the brain have been called out as a basis for consciousness itself.[7]

1 Wallin, N. L., B. Merker, and S. Brown. *The Origins of Music*. Cambridge, MA: MIT P, 2000.

2 Lord, A. B. *The Singer of Tales*. Harvard Studies in Comparative Literature. Harvard UP, 1960.

3 Gioia, T. *Work Songs*. Durham, NC: Duke UP, 2006.

4 Pham, Ha. "West Africa Ghana, Post Office." www.Youtube.Com/Watch?V=C3fctmixske. YouTube. 2011. Web.

5 *The Woven Sounds (Documentary Film)*. 2019. Documentary.

6 Brown, S., and J. Jordania. "Universals in the World's Musics." *Psychology of Music*, vol. 41, no. 2, 2011, 229–48.

7 Dehaene, S. *Consciousness and the Brain: Deciphering How the Brain Codes Our Thoughts*. New York, Viking, 2014.

Language probably does *not* immediately come to mind when we think of rhythm. You might have had a high school literature class where you learned about prosodic feet—iambs, trochees, and anapests. But outside the context of poetry, we rarely think about speech having a particular rhythm. After all, we are likely to say "Oy Bill—you ready yet?" Not, "Hey there Bill, / do you think / it is now / time to go?" so that it conforms to dactylic tetrameter. What about rhythm and reading? Here, too, we are unlikely to associate rhythm to reading unless we are reading poetry. In fact, *rhythm is a necessary ingredient of linguistic communication itself.*

Rhythms Fast and Slow

Rhythm can be viewed through the lens of shorter and longer timescales. Speech has phoneme-, syllable-, word-, and sentence-length rhythmic units, each unfolding at their own rate. We understand that speech comes in different sized units—the sound an individual letter makes, the phoneme, at one extreme, and the slowly rising and falling loudness and pitch contours that unfold over the course of a sentence or group of thoughts on the other. This latter one is the nighttime reading rhythm I fall asleep to. These entwined elements of speech constitute rhythms that must be sorted by our sound minds, my term for the vast resources our brain devotes to the processing of sound and how our brain, in turn, impacts our sonic world.[8] We can try to focus on the slow parts of speech (say, the fluctuating pitch of the voice) and ignore the fast (the vowel and consonant sounds that convey the meaning of the words), or vice versa. But this is usually not possible and rarely desirable.

This temporal hierarchy is at work in music too. Music is a mix of slow phrases, steady beats, sustained notes, rapidly changing notes, trills, and drum crashes. Entwined temporal structures are in environmental sounds as well—when walking through the woods, we simultaneously hear slow footsteps, the unfolding crunch of leaves underfoot, and the rapid snap of a twig. Much as sound units come in different lengths, brain rhythms come in different speeds. Subcortical structures are equipped for microsecond timing while the cortex is better suited to integrating sounds over a longer timescale.

[8] Kraus, N. *Of Sound Mind: How Our Brain Constructs a Meaningful Sonic World.* MIT P, 2021.

Brain rhythms can be measured both when at rest and when performing an activity. When listening to speech, there are fast brain rhythms that entrain to the fast phonemes, the near instantaneous consonant sounds. Middle-range rhythms in the brain track the rate of syllables. Slower brain rhythms correspond to the slow oscillations of phrases and sentences.[†,9] Similar nested brain patterns are active when listening to music (Figure 2.1).

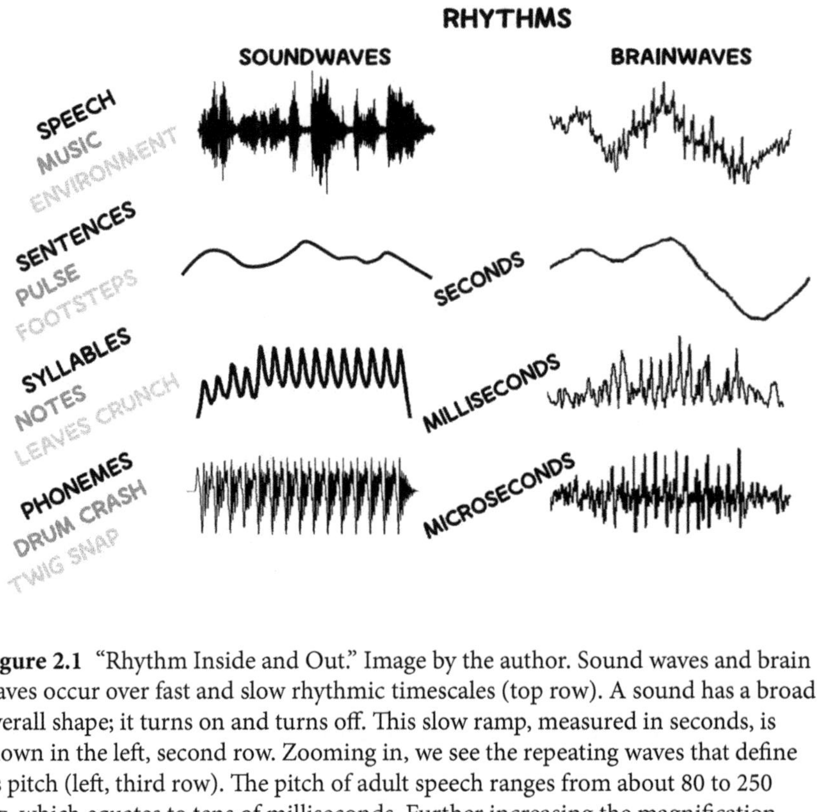

Figure 2.1 "Rhythm Inside and Out." Image by the author. Sound waves and brain waves occur over fast and slow rhythmic timescales (top row). A sound has a broad overall shape; it turns on and turns off. This slow ramp, measured in seconds, is shown in the left, second row. Zooming in, we see the repeating waves that define its pitch (left, third row). The pitch of adult speech ranges from about 80 to 250 Hz, which equates to tens of milliseconds. Further increasing the magnification are the vowels and consonants with frequencies up to the thousands of Hz, or in the microsecond timescale (left, fourth row). Each of these timescales (seconds, milliseconds, microseconds) is maintained simultaneously in the brain (right column).

† These brain rhythms are named with Greek letters. Exact frequency cutoffs are somewhat fluid, but roughly, the slowest are delta (1–4 Hz) and theta (4–8 Hz) and the fastest is gamma (30–70 Hz), with alpha and beta falling in between. These ranges span sentence-slow to phoneme-fast.

9 Kotz, S. A., A. Ravignani, and W. T. Fitch. "The Evolution of Rhythm Processing." *Trends in Cognitive Sciences*, vol. 22, no. 10, 2018, 896–910.

Rhythm is *in* Us

We've all heard him. The beginning piano player. The "Eensie Weensie Spider" and "Frère Jacques" plinking hesitantly out of the piano. Usually getting the notes right is paramount to a beginner; hitting the right keys trumps hitting them at the right time. It is heartwarming to listen as our child plays (mostly) correct notes at wildly incorrect times. What goes on in the brain when we hear music being played on beat or off beat?

Imagine a metronome ticking at about 144 beats per minute (bpm). Popular songs in this range include Blondie's "Call Me," the Beatles' "Back in the USSR," and the Rolling Stones' "(I Can't Get No) Satisfaction." It's a fast, *allegro* rate. Measured another way, these songs have about a half second between their beats. If we play a conga drum by itself at this rate and record brain waves to it, we will see neural activity repeating every half second (boom, boom, boom, boom, or "one, two, three, four"). But what does the brain do if you listen to the conga drumming along to a song that matches this beat? The brain produces a new rhythm! In addition to a response peak every half second (where musically speaking the "ones" are), you see another, smaller peak halfway in between (1 *and* 2 *and* 3 *and* 4 *and*; "FLEW in FROM mi-AM-I BEACH"). The brain has worked out the strong/weak pairs comprising the song's meter. This tells us the brain entrains and reinforces both explicit and implied rhythms in the music.[10] This extra rhythm in the brain wave does not occur when the song is deliberately misaligned with the conga beat. A similar example of the brain creating a beat comes from a study showing that the fundamental frequency of an identical speech sound is enhanced when it occurs on the "1" in a four-beat sequence.[11] The sound mind's response to a drumbeat is deeply shaped by its aural context. Rhythmic organization operates automatically when we listen to sound. If our rhythmic expectations are violated, our brains behave in a different manner because of our inherent internal sense of rhythm.

10 Tierney, A., and N. Kraus. "Neural Entrainment to the Rhythmic Structure of Music." *Journal of Cognitive Neuroscience*, vol. 27, no. 2, 2015, 400–408.

11 Moon, I. J., et al. "Meter Enhances the Subcortical Processing of Speech Sounds at a Strong Beat." *Scientific Reports*, vol. 10, no. 1, 2020, 15973.

Rhythm Intelligences

Imagine the familiar rhythm "Shave and a haircut, two bits" and tap it out on a table with your finger. Did you tap seven times? Now imagine it again and tap your foot to it. Did you tap seven times again? Or fewer? For me, when I tap my finger on a table, I tap to every note (ignoring the rests). When I tap my foot or snap my finger along to music, I typically tap or snap to the beat (or pulse) of the song, not every note. When I tap my finger on the table, hitting the "sounds" and ignoring the "silences," I am tapping out the *rhythmic pattern*—I am keeping track of how long or short each note is and where the pauses occur. When I tap my foot, I tap four times, to the underlying beat or pulse (Figure 2.2), which in this example includes a silent beat. Music has both a pulse and a rhythm pattern, notated by time signature and note/rest durations, respectively.[12]

Before I began studying rhythm, if you had asked me about the skills involved in tapping out rhythmic patterns versus tapping out the beat, I would have said, "You're probably either good at both or not so good at either." If someone can tap to a beat, they can also tap out a rhythm pattern, right?

Wrong. There are *multiple rhythmic intelligences*. You cannot predict how someone will perform one rhythm task by how they perform a different rhythm task. This was first noticed in extreme cases where a person with brain damage could be impaired in one sort of rhythmic ability but not

Figure 2.2 "Rhythm Inside and Out." Image by the author. The rhythm pattern of "Shave and a haircut, two bits" is defined by the duration of the notes and rests that make up the tune (top row of arrows). In musical notation, pulse is signified by the time signature, in this case 4/4. The bottom row of arrows depicts the four beats, which may occur during either a note or a rest. Can you simultaneously tap your foot to the beat and your finger to the rhythm pattern?

[12] Kraus, N. "Rhythm: A Case for Digital Music Medicine." *Hearing Journal*, vol. 74, no. 7, 2021, 6–9.

another.[13] We have since learned these distinctions are fundamental to how the system works: we see dissociations between rhythm skills in all of us,[14] confirming the idea that "rhythm" is not an all-or-nothing ability, and more intriguingly, our proficiency executing one type of rhythm or another bears on various real-life skills. Specifically, beat-keeping aligns with reading, attention, linguistic, and inhibitory control skills, whereas a facility for producing rhythm patterns aligns with memory and the all-important ability to understand speech in noise.[15]

Brain Rhythms

Rhythm pattern skills are associated with slower brain rhythms (seconds), whereas beat-keeping skills are associated with faster brain rhythms (milliseconds and microseconds).[16] Phonemes, syllables, and sentences range from microsecond to millisecond to second timing, respectively. Brain rhythms can predict language development in infants and children.[17] Brain rhythms can also determine a person's strengths and bottlenecks related to language and the ability to make sense of an auditory scene while listening in noise.

[13] Fries, W., and A. A. Swihart. "Disturbance of Rhythm Sense Following Right Hemisphere Damage." *Neuropsychologia*, vol. 28, no. 12, 1990, 1317–323.
Di Pietro, M., et al. "Receptive Amusia: Temporal Auditory Processing Deficit in a Professional Musician Following a Left Temporo-Parietal Lesion." *Neuropsychologia*, vol. 42, no. 7, 2004, 868–77.
Peretz, I. "Processing of Local and Global Musical Information by Unilateral Brain-Damaged Patients." *Brain*, vol. 113, no. 4, 1990, 1185–205.
Li.geois-Chauvel, C., et al. "Contribution of Different Cortical Areas in the Temporal Lobes to Music Processing." *Brain*, vol. 121, no. 10, 1998, 1853–67.

[14] Tierney, A., and N. Kraus. "Evidence for Multiple Rhythmic Skills." *PLoS One* 10.9 (2015): e0136645
Bonacina, S., et al. "How Rhythmic Skills Relate and Develop in School-Age Children." *Global Pediatric Health*, vol. 6, 2019, 1–7.

[15] Tierney, A., et al. "Individual Differences in Rhythm Skills: Links with Neural Consistency and Linguistic Ability." *Journal of Cognitive Neuroscience*, vol. 19, no. 5, 2017, 855–68.
Thomson, J. M., and U. Goswami. "Rhythmic Processing in Children with Developmental Dyslexia: Auditory and Motor Rhythms Link to Reading and Spelling." *Journal of Physiology*, vol. 102, nos. 1–3, 2008, 120–29.
Bonacina, S., et al. "Clapping in Time Parallels Literacy and Calls Upon Overlapping Neural Mechanisms in Early Readers." *Annals of the New York Academy of Sciences*, vol. 1423, 2018, 338–48
Slater, J., et al. "Speech-in-Noise Perception Is Linked to Rhythm Production Skills in Adult Percussionists and Non-Musicians." *Language, Cognition and Neuroscience*, vol. 33, no. 6, 2018, 710–17.

[16] Tierney, A., and N. Kraus. "Getting Back on the Beat: Links between Auditory–Motor Integration and Precise Auditory Processing at Fast Time Scales." *European Journal of Neuroscience*, vol. 43, no. 6, 2016, 782–91.

[17] Benasich, A. A., et al. "Early Cognitive and Language Skills Are Linked to Resting Frontal Gamma Power across the First 3 Years." *Behavioural Brain Research*, vol. 195, no. 2, 2008, 215–22.

Rhythm, Language, Listening, and Memory

Rhythm is tied to language. Children who recognize differences in rhythm patterns and tap to a beat learn to read and spell more easily.[20,18,19,20,21,22] Several beat-keeping skills are impaired in older children with dyslexia.[25] We have found a link between beat keeping and language development in adolescents[28] and in children as young as 3 years old.[21,23,24] What is the connection between rhythm skills and what might appear to be unrelated skills like reading and writing?

There really is rhythm in language, beyond the rhyming of poetry. It is inherently a part of pronunciation. Rhythm matters, even in single words. "Record," "contrast," "project," and "produce" can be either nouns or verbs depending on which syllable is stressed. Running speech also has rhythm. A YouTube search for "drumming to speech" will uncover some nice examples; a personal favorite is the one with the scene from the Gene Wilder *Willy Wonka* movie. The video shows a drummer playing along to the rhythms of the dialogue between Willy and Grandpa Joe, so you cannot miss the rhythm in speech. Tabla player Zakir Hussain tells us his father taught him to speak using drum rhythms when he was a baby. In tabla, each finger is assigned a syllable, and playing the tabla is akin to speaking in phrases. In all languages there is a definite rhythmic aspect to spoken language, brought about by alterations in stress, duration, and pitch of the syllables. This was resoundingly brought home to me firsthand when Zakir accompanied me on the congas during a speech on rhythm and language. Quite simply, rhythm in speech tells us when important information starts and stops. Stressed syllables emerge at roughly regular intervals and, importantly, carry the majority of the information of speech. With an ongoing rhythmic flow, the listener is guided to the important features of the sentence by the expectancy

[18] Wolff, P. "Timing Precision and Rhythm in Developmental Dyslexia." *Reading Writing* 15 (2002): 179–206.

[19] Thomson, J., et al. "Auditory and Motor Rhythm Awareness in Adults with Dyslexia." *J Res Reading* 29 (2006): 334–48.

[20] Corriveau, K. H., and U. Goswami. "Rhythmic Motor Entrainment in Children with Speech and Language Impairments: Tapping to the Beat." *Cortex* 45.1 (2009): 119–30.

[21] Tierney, A. T., and N. Kraus. "The Ability to Tap to a Beat Relates to Cognitive, Linguistic, and Perceptual Skills." *Brain Lang* 124.3 (2013): 225–31.

[22] Moritz, C., et al. "Links between Early Rhythm Skills, Musical Training, and Phonological Awareness." *Reading Writing* 26 (2013): 739–69.

[23] Woodruff Carr, K., et al. "Beat Synchronization Predicts Neural Speech Encoding and Reading Readiness in Preschoolers." *Proc Natl Acad Sci U S A* 111.40 (2014): 14559–64.

[24] Bonacina, S., et al. "Distinct Rhythmic Abilities Align with Phonological Awareness and Rapid Naming in School-Age Children." *Cogn Process* 21 (2020): 575–81.

rhythm sets up and, so primed, we understand the content of the spoken word better.[25] With the understanding of spoken speech comes the ability, when learning to read, to make the necessary connections between the sounds of language and its written form.

One of the greatest impediments to successful spoken-word communication is noise. The rhythm of speech helps us. This is because the rhythm in speech helps us fill in the gaps when noise causes us to miss a few words. Just as a rhythm pattern evolves over the course of a measure of music, running speech evolves over time and thus is suited for a slower scale of auditory processing. The strong and weak stresses, phrases, and boundaries of language are relevant to the whole spoken sequence. The ability to reproduce rhythm patterns seems to draw on the same skills required for forming the auditory scenes that comprise hearing speech that is barely audible above noise.

Hearing speech in noise can be partly predicted by one's ability to tap out rhythm patterns.[22] And the better you are at navigating rhythms—and musicians of any stripe (not just drummers) fall into this category—the more you can capitalize on the rhythmic patterns in speech and eke out what was said despite the noise.[26,27,28,29,30,31,32,33]

The common denominator between facility at rhythm pattern production and listening in noise is probably working memory. Memory for the sounds demarcating rhythm in music, and more broadly all sounds, is likely stored in sensory neurons themselves. Biologically speaking, my framework for learning through sound, the *BEAMS* hypothesis,[34] posits how the sound mind interacts

[25] Kohler, K. J. "Rhythm in Speech and Language: A New Research Paradigm." *Phonetica* 66.1–2 (2009): 29–45.

[26] Kraus, N., and T. White-Schwoch. "Neurobiology of Everyday Communication: What Have We Learned from Music?" *Neuroscientist* 23.3 (2017): 287–98.

[27] Parbery-Clark, A., et al. "Musician Enhancement for Speech-in-Noise." *Ear and Hearing* 30.6 (2009): 653–61.

[28] Parbery-Clark, A., et al. "Musical Experience and the Aging Auditory System: Implications for Cognitive Abilities and Hearing Speech in Noise." *PLoS One* 6.5 (2011): e18082.

[29] Parbery-Clark, A., et al. "Musicians Have Fine-Tuned Neural Distinction of Speech Syllables." *Neuroscience* 219 (2012): 111–19.

[30] Zendel, B. R., and C. Alain. "Musicians Experience Less Age-Related Decline in Central Auditory Processing." *Psychol Aging* 27.2 (2012): 410–7.

[31] Strait, D. L., et al. "Musical Training During Early Childhood Enhances the Neural Encoding of Speech in Noise." *Brain Lang* 123.3 (2012): 191–201.

[32] Swaminathan, J., et al. "Musical Training, Individual Differences and the Cocktail Party Problem." *Sci Rep* 5 (2015): 11628.

[33] Zendel, B. R., et al. "The Impact of Musicianship on the Cortical Mechanisms Related to Separating Speech from Background Noise." *J Cogn Neurosci* 27.5 (2015): 1044–59.

[34] Kraus, N. "Memory for Sound: The Beams Hypothesis [Perspective]." *Hear Res* 407 (2021): 108291.

with the other senses, and the motor, emotion, cognitive, and visceral systems of the *Brain*. It describes how experience with sound first effects changes through *Efferent* (brain-to-ear) neural circuitry. Over time, as the importance of a given sound is established, *Afferent* (ear-to-brain) pathways change, representing *Memories* of the *Sounds* of our lives. Indeed, the auditory and motor pathways involved in rhythm processing are retuned to reflect experience with rhythm and positively impact our communication abilities.

Rhythm and Vocal Learning

Have you heard about Snowball? If you haven't, stop reading right now and search for "Snowball the cockatoo" on YouTube. Snowball the sulfur-crested cockatoo dances along to pop music. There is no mistaking it. Snowball bobs his head and steps his feet in time to the rhythm of Michael Jackson, Lady Gaga, and, most famously, the Backstreet Boys. John Iversen and Ani Patel studied Snowball's dancing.[35] By systematically varying tempos and observing corresponding shifts in Snowball's movement, they verified that Snowball was reacting to the beat. This is in contrast to "dancing" horses, for example, who do not react to music but whose rhythmic steps are cued by the rider. So Snowball's dancing, in addition to being seriously cool, raises some questions. Who else can do this? Can other bird species dance? Any other animals? Why can't my dog—he's smart? Chimps must be able to, right? They are closer to humans than a cockatoo. It turns out Snowball is part of a small and select group of animals that can keep a beat. To date, beat keeping is confirmed only in a variety of birds including parrots and cockatoos, seals, sea lions, elephants, rats, and humans. That's it.

What does this seemingly disparate group of animals have in common? Along with bats, whales, seals, hummingbirds, and songbirds, most are "vocal learners." This means these species have the ability to imitate new sounds they hear. Most animals, however intelligent, cannot imitate sounds. Take your dog, for example. She may know a dozen or more words. I have never met a dog who does not understand the word "walk." And yet, however deep the associations in her mind between certain words and what they represent, she will never *say*

[35] Patel, A. D., et al. "Experimental Evidence for Synchronization to a Musical Beat in a Nonhuman Animal." *Curr Biol* 19.10 (2009): 827–30.

"walk." Dogs, along with most of the animal kingdom, are limited to a handful of vocalizations. Vocal learners, however, can move beyond their innate sounds. Parrots can "speak." Songbirds learn their songs via imitation; a bird raised away from others of his species will not develop the same song as his peers but rather an impoverished and unstructured aberrant deviation. Human speech, of course, is a testament to our species being masters of vocal learning. These imitative abilities arise from extensive connections in the brain between the auditory and motor regions that are missing in most other species. A by-product of this connectivity is the ability *to predict the timing of future beats*. This is the key to Snowball's and humans' beat keeping. We do not simply react, like a horse or a chimpanzee, to a present or past cue; we anticipate future beats and can move our bodies accordingly.

Rhythm and Movement

Sound is motion, the motion of air. We have spoken about rhythm thus far as something that is heard. However, the flip side is movement. You cannot hear a drum unless someone beats it. You cannot snap your fingers along to a song unless you, well, move your fingers. You cannot produce speech without moving your mouth. Moving is intertwined with hearing when it comes to making and listening to music and speech. When just *listening* to speech, or even *imagining* a singer, the brain areas devoted to the movement of our mouths are activated.[36,37] Likewise, listening to a piano melody activates the parts of the motor system involved in controlling fingers in people who know how to play the tune on the piano.[38] Your sound mind "moves" to music even if you are quiet as a statue, especially if it is a piece of music you have played.

If you are walking down the street talking to a friend, it is likely the two of you unconsciously synchronize your footsteps.[39] This synchronization helps us communicate—the number of footfalls is halved, thus decreasing the chance

[36] Wilson, S. M., et al. "Listening to Speech Activates Motor Areas Involved in Speech Production." *Nat Neurosci* 7.7 (2004): 701–2.

[37] Herholz, S. C., et al. "Dissociation of Neural Networks for Predisposition and for Training-Related Plasticity in Auditory–Motor Learning." *Cereb Cortex* 26.7 (2016): 3125–34.

[38] Bangert, M., et al. "Shared Networks for Auditory and Motor Processing in Professional Pianists: Evidence from fMRI Conjunction." *Neuroimage* 30.3 (2006): 917–26.

[39] Larsson, M., S.R. Ekstrom, and P. Ranjbar. "Effects of Sounds of Locomotion on Speech Perception." *Noise Health* 17.77 (2015): 227–32.

that any given speech sound is drowned out by a footstep. Not only can you hear your friend better but, if you and your friend happen to be a pair of wild animals, you are better able to monitor and detect nearby prey or predators.

Infants just days old attend to rhythms,[40] but what determines which rhythms they choose to listen to? It turns out the motor component of rhythm plays a role in preferences. In one study, 7-month-old babies were exposed to an ambiguous rhythm pattern.[41] The ambiguity arose because one could ascribe either a 2/4 or a 3/4 time signature to it. That is, you could reasonably count out either **1**-2-**1**-2-**1**-2 or **1**-2-3-**1**-2-3. Experimenters bounced babies at one of the two possible meters—baby May was bounced on every second beat and baby June was bounced on every third beat. Then later, without being bounced, each heard *stressed* versions of the same two rhythms, meaning this time the location of the 1's was obvious. May preferred the 2/4 stressed one and June preferred the 3/4 one, as measured by how long they listened to each before turning away. Rhythm preferences are formed early in life. Notably, babies did not form a preference for 2/4 or a 3/4 meter when they simply observed someone else bouncing; their own bodies had to move to seal the deal.

Rhythm and Socialization

How we feel about another person is conveyed by rhythm. Walkers synchronize their steps to aid communication. Snowball will dance along with you, but if you dance out of sync with the beat, he will turn away from you. Social encounters with rhythm influence our attitudes. The extent to which a person synchronizes with an experimenter affects the person's opinion of the likability of the experimenter. University students were instructed to tap along to a metronome while an experimenter was also tapping their finger nearby. When the experimenter tapped at the same rate, the rating given in response to "how likable was the experimenter?" was higher.[42] Likability aside, the mere presence of another person drumming along with a task will improve performance.

[40] Winkler, I., et al. "Newborn Infants Detect the Beat in Music." *Proc Natl Acad Sci U S A* 106.7 (2009): 2468–71.

[41] Phillips-Silver, J., and L. J. Trainor. "Feeling the Beat: Movement Influences Infant Rhythm Perception." *Science* 308.5727 (2005): 1430.

[42] Hove, M. J., and J. L. Risen. "It's All in the Timing: Interpersonal Synchrony Increases Affiliation." *Soc Cogn* 27.6 (2009): 949–61.

Preschool-age children asked to perform a rhythm synchronization task perform it better if they are drumming along with another human than to an impersonal beat coming from a loudspeaker.[43]

Even in very young children, being (literally) "in sync" with another person engenders positive feelings toward them. An experimenter bounced 14-month-old children along to music either on the beat or intentionally off the beat. When the bouncing session was over, the baby was placed on the floor and the experimenter deliberately dropped an object and acted out needing help to pick it up. The babies who were bounced on-beat were much more likely to help the experimenter retrieve the object, having apparently formed a social bond, via rhythm, that prompted cooperation. The off-beat babies were less likely to help.[44] *Rhythmic* synchrony had led to *interpersonal* synchrony.

Along similar lines, the brain rhythms of musical performers and their audiences have been measured in concert settings. The brain rhythms tend to synchronize, and the more synchronization between performer and listener, the more listeners report enjoying the performance.[45] Other physiological metrics such as heartrate and respiration align in audience members listening to music, and the interperson synchrony is greatest when listeners feel moved or inspired by the music[46] or to passages considered structurally important.[47]

Music in general, and rhythm in particular, does an uncommonly good job fostering a sense of community.[48] Indeed, music being played at negotiation sessions helps to smooth the conversations and leads to breakthroughs and compromises. Musicians Without Borders is used to form relationships in troubled regions around the world, to bring hope, comfort, and healing to diverse populations.[49] The Resonance Project and the Jerusalem Youth Chorus, which are forming bonds between Israeli and Palestinian children, are other examples of using musical rhythm to overcome differences. The early days of

[43] Kirschner, S., and M. Tomasello. "Joint Drumming: Social Context Facilitates Synchronization in Preschool Children." *J Exp Child Psychol* 102.3 (2009): 299–314.

[44] Cirelli, L. K., K. M. Einarson, and L. J. Trainor. "Interpersonal Synchrony Increases Prosocial Behavior in Infants." *Developmental Science* 17.6 (2014): 1003–11.

[45] Hou, Y., et al. "The Averaged Inter-Brain Coherence between the Audience and a Violinist Predicts the Popularity of Violin Performance." *Neuroimage* 211 (2020): 116655.

[46] Tschacher, W., et al. "Audience Synchronies in Live Concerts Illustrate the Embodiment of Music Experience." *Sci Rep* 13.1 (2023): 14843.

[47] Czepiel, A., et al. "Synchrony in the Periphery: Inter-Subject Correlation of Physiological Responses During Live Music Concerts." *Sci Rep* 11.1 (2021): 22457.

[48] Kraus, N. "Sound Connects Us." *Music and Mind: Harnessing the Arts for Health and Wellness.* Edited by Fleming, Renée. New York, NY: Viking, 2024, pp. 38–56.

[49] "Musicians without Borders." Web2020. www.musicianswithoutborders.org.

the 2020 coronavirus pandemic were marked, in some European countries, by daily sessions of songs sung from balconies to connect with others at a time of isolation and to communicate appreciation and solidarity with healthcare workers.

Rhythm for Health

Traditional healers in all regions of the world have relied on rhythm as a primary force in their rituals and practice.[50] Today, rhythm helps us exercise as we move to keep ourselves healthy.[51] Therapists have long used our capacity to perceive sound patterns to strengthen communication skills. They rely on rhythm and the concepts of entrainment to a beat, violations of a beat, and pattern recognition as core features of their protocols,[52] reminiscent of the scene in the Colin Firth film *The King's Speech*, where King George VI overcame a stuttering problem by rhythmically singing his words. Rhythm capitalizes on our sound minds' auditory–motor connection.

First mentioned by the American Medical Association in 1914, music therapy was put to work helping wounded World War soldiers recover from their injuries, including what we now call traumatic brain injury. Rhythm skills are disrupted in children with concussions[53] and rhythm-based therapy has a growing status in recovery from concussion and other brain injuries, addressing both cognitive and emotional health.[54,55,56] Rhythm is used to great effect to pace walking in

[50] Gioia, T. *Healing Songs*. Durham, NC: Duke UP, 2006.

[51] Waterhouse, J., P. Hudson, and B. Edwards. "Effects of Music Tempo Upon Submaximal Cycling Performance." *Scand J Med Sci Sports* 20.4 (2010): 662–9.

[52] Lim, H. A. "Effect of 'Developmental Speech and Language Training through Music' on Speech Production in Children with Autism Spectrum Disorders." *J Music Ther* 47.1 (2010): 2–26.

[53] Bonacina, S., et al. "Concussion Disrupts Brain Synchrony: Evidence from Interactive Metronome on Young Children with Persisting Symptoms and Prolonged Recovery Post-Concussion." *Journal of Concussion* 7 (2023): 1–5.

[54] Nelson, L. A., et al. "Effects of Interactive Metronome Therapy on Cognitive Functioning after Blast-Related Brain Injury: A Randomized Controlled Pilot Trial." *Neuropsychology* 27.6 (2013): 666–79.

[55] Hegde, S. "Music-Based Cognitive Remediation Therapy for Patients with Traumatic Brain Injury." *Front Neurol* 5 (2014): 34.

[56] Thaut, M. H., et al. "Neurologic Music Therapy Improves Executive Function and Emotional Adjustment in Traumatic Brain Injury Rehabilitation." *Ann N Y Acad Sci* 1169 (2009): 406–16.

individuals with movement disorders such as Parkinson's disease.[57,58,59] After all, walking *is* a rhythm. Other disorders that involve movement, such as aphasia, stuttering, difficulty with respiration, swallowing, and speaking, respond to music therapy.[60,61,62,63,64,65,66]

Therapy involving rhythm also has shown promise in addressing communication and social behavior in people on the autism spectrum.[67,68,69] Children who cannot otherwise speak can form words and sentences when accompanied by a clear rhythm. There are children on the autism spectrum who will not engage in a verbal conversation but will gladly carry on a *rhythmic* conversation with another person on drums. And, moving in synchrony positively affects how we feel about each other.[51]

If I had a magic wand, I would make rhythm an indisputable part of language therapies through music- and rhythm-based instruction. This would mean a closer alignment among fields of speech therapy, music, and music therapy. There are explicit rhythm-based training programs that make synchronizing to rhythm an explicit core exercise with the aim of improving timing in the brain. Some have been used to bolster language, reading, and communication skills,

57 Nombela, C., et al. "Into the Groove: Can Rhythm Influence Parkinson's Disease?" *Neurosci Biobehav Rev* 37.10 Pt 2 (2013): 2564–70.

58 de Dreu, M. J., et al. "Rehabilitation, Exercise Therapy and Music in Patients with Parkinson's Disease: A Meta-Analysis of the Effects of Music-Based Movement Therapy on Walking Ability, Balance and Quality of Life." *Parkinsonism Relat Disord* 18 Suppl 1 (2012): S114–9.

59 Hausdorff, J. M., et al. "Rhythmic Auditory Stimulation Modulates Gait Variability in Parkinson's Disease." *Eur J Neurosci* 26.8 (2007): 2369–75.

60 Tomaino, C. M. "Recovery of Fluent Speech through a Musician's Use of Prelearned Song Repertoire: A Case Study." *Music and Medicine* 2.2 (2010): 85–88.

61 Tomaino, C. M "Effective Music Therapy Techniques in the Treatment of Nonfluent Aphasia." *Annals of the New York Academy of Sciences* 1252.1 (2012): 312–7.

62 Stegemoller, E. L., et al. "Experiences of Persons with Parkinson's Disease Engaged in Group Therapeutic Singing." *J Music Ther* 54.4 (2018): 405–31.

63 Raglio, A., et al. "Improvement of Spontaneous Language in Stroke Patients with Chronic Aphasia Treated with Music Therapy: A Randomized Controlled Trial." *Int J Neurosci* 126.3 (2016): 235–42.

64 Thaut, M. H., and G. C. McIntosh. "Neurologic Music Therapy in Stroke Rehabilitation." *Current Physical Medicine and Rehabilitation Reports* 2.2 (2014): 106–13.

65 Tomaino, C. M. "Clinical Applications of Music Therapy in Neurologic Rehabilitation." *Music That Works*. Edited by Haas, R., and V. Brandes CVVinvVVVVienna: Springer-Verlag, 2009, pp. 211–20.

66 Brady, J. P. "Metronome-Conditioned Speech Retraining for Stuttering." *Behavior Therapy* 2.2 (1971): 129–50.

67 Hardy, M. W., and A. B. Lagasse. "Rhythm, Movement, and Autism: Using Rhythmic Rehabilitation Research as a Model for Autism." *Front Integr Neurosci* 7 (2013): 19.

68 LaGasse, A. B. "Effects of a Music Therapy Group Intervention on Enhancing Social Skills in Children with Autism." *J Music Ther* 51.3 (2014): 250–75.

69 LaGasse, A. B. "Social Outcomes in Children with Autism Spectrum Disorder: A Review of Music Therapy Outcomes." *Patient Relat Outcome Meas* 8 (2017): 23–32.

and do so with tasks that engage both the slow and fast sound processing circuits in the brain, thus drawing on multiple rhythm intelligences.[21,70,71,72]

Music with a regular and predictable rhythm can lead to states of enjoyment or emotional transcendence.[73] Pythagoras viewed music as a gateway to the realm of the dead, at least judging from his supposed dying request that the monochord, an ancient one-stringed instrument, be played during his final moments. Gregorian chants have been described as "so rich in overtones that you have the impression they are angels, not men."[57] Grateful Dead drummer Mickey Hart and I have discussed the calm yet alert and energized state that drone compositions—musical pieces consisting of sustained sounds produced by monochords or other instruments and manipulated in the studio to swell and build—can induce. We are working together to investigate the neurophysiological reaction to some of his drone compositions.

A while back, one of my sons got a hairline fracture in his foot. As he was not healing as fast as his physical therapist was hoping, he was assigned daily sessions with a bone vibrator. The idea behind vibration therapy is that if you cannot use your musculoskeletal system normally, for example due to an injury or osteoporosis, you miss out on the natural stimulation that occurs as your muscles imperceptibly relax and contract to maintain posture. This can lead to bone tissue atrophy. Imposing vibrations at around 30–50 Hz at the injury site simulates natural postural adjustments, stops the reabsorption of bone tissue, and promotes the bone growth that would ordinarily be achieved as part of typical day-to-day movement.[74] It appears that low-frequency vibrations spur activity in the stem cells that make cartilage, muscles, and bones. This process may also be useful for strength training in noninjured people.

It turns out the vibration rate of a cat purr is in the exact same range as used in vibration therapy for bone growth. Cats purr when they're happy, of course, but in what other circumstance do they purr? When they are injured! There is a hypothesis that cats purr as a mechanism to keep their bones and muscles

[70] Ritter, M., K. A. Colson, and J. Park. "Reading Intervention Using Interactive Metronome in Children with Language and Reading Impairment: A Preliminary Investigation." *Communication Disorders Quarterly* 34.2 (2013): 106–19.

[71] Taub, G. E., K. S. McGrew, and T. Z. Keith. "Improvements in Interval Time Tracking and Effects on Reading Achievement." *Psychology in the Schools* 44.8 (2007): 849–63.

[72] Bonacina, S., et al. "Clapping in Time with Feedback Relates Pervasively with Other Rhythmic Skills of Adolescents and Young Adults." *Percept Mot Skills* 128.3 (2021): 952–68.

[73] Barrett, F. S., et al. "Qualitative and Quantitative Features of Music Reported to Support Peak Mystical Experiences During Psychedelic Therapy Sessions." *Front Psychol* 8 (2017): 1238.

[74] Thompson, W. R., S. S. Yen, and J. Rubin. "Vibration Therapy: Clinical Applications in Bone." *Curr Opin Endocrinol Diabetes Obes* 21.6 (2014): 447–53.

stimulated and healthy and to restore their health when injured.[75] Maybe it is not a coincidence that cats have better bone health and a lower incidence of osteoporosis than dogs. Maybe this is the secret to their nine lives.

Conclusion

Why do we care about rhythm? Because *sound is motion and sound moves us*. The auditory and motor systems combine to enable us to communicate. When we engage with rhythm, whether a precise in-the-moment beat or a longer rhythm sequence, we rely on precise timing. Our very brains are wired for rhythm across these timescales. The currency of the nervous system—electricity—is nothing if not rhythmic. To my mind, the crackling of action potentials as they respond to sound is closer to a one-to-one correspondence of stimulus to response than in any other sensory modality. Physiologists commonly capitalize on this and literally "listen to the brain" to help guide their electrode placement during their experiments by playing the crackle of neural activity through a speaker. I love to listen to the brain's language of timed, rhythmic (and arrhythmic) electrical impulses. The better we understand the biological basis of rhythm, the better we will be able to employ rhythm—in all its guises—to improve communication and to better understand ourselves.

[75] Muggenthaler, E. "The Felid Purr: A Healing Mechanism?" *The Journal of the Acoustical Society of America* 110 (2001): 2666.

3

Rhythmic Operations as Boundary Conditions

Mark Lussier

> Everywhere where there is interaction between a place, a time, and an expenditure of energy, there is rhythm.
>
> (Henri Lefebvre, *Elements of Rhythmanalysis* 15)

> Meter is dogmatic, but rhythm is critical; it ties together critical moments or ties itself together in passing from one milieu to another. [...] Rhythm is never on the same plane as that which has rhythm.
>
> (Gilles Deleuze and Felix Guattari, *A Thousand Plateaus* 313)

> Work to the rhythm, Live to the rhythm, Love to the rhythm, Slave to the rhythm.
>
> (Grace Jones, *Slave to the Rhythm*)

I

The lyrics of Grace Jones's best-known song, *Slave to the Rhythm*, catalogue the omnipresent and tyrannical aspects of "the rhythm" (a force emphasized by the article "the"), which functions as an oscillatory background radiation (to borrow an analogy from physics) with which work, life, and love interface and interfere. The forms of rhythm evoked in the song both assert and embody this universal constant, a lyrical propagation supplied primarily by the linguistic pulse within the verse, the repetition of the lower-case phrase "to the rhythm." The musical trajectory of the song, in partnership with its linguistic elements, stages a collision among diverse formal elements, providing creative opportunities to experience a rhythm revealed/unveiled while seeking a rhythm unbound. Literary and musical studies have long investigated the capacity of songs to entangle notes and words—two primary forms of rhythm—to achieve "phenomenological" (i.e., physiological and psychological) effects (Patel 342) that help understand "the entangling rhythms that circulate in and outside the body," which thereby prepares auditors for a confrontation with "the complex array of non-human rhythms that impose upon, exist separately [yet] are entangled with human rhythms" (Edensor 5, 7).

As Janina Wellman observes, "the regularity in the alternation of note values—rhythm—is the autonomy of the movement," and the song's lyrics and metrics inevitably implicate all who encounter them via acts of participation, reception and response to "this autonomy of the movement" (64), especially in light of the expansive and inclusive categories of work, life, and love previously referenced in the lyrics, activities that often define "the rhythm/s of social experience" (Levine 51). These practices and processes, although part circadian and part constructed, remain unvaryingly governed by rhythmic patterns of interference in body and mind (as if these were separate), engaging an encompassing world of rhythmic operations (from top to bottom, side to side, "Everything Everywhere All at Once" [*c.*2024]). Amittai Aviram re/cognized thirty years ago that "both cognitive apprehension [...] and physical involvement or participation" (43) implicate all observers as sentient and participatory ripples within the surrounding vibrant threads of Indra's net, a Buddhist concept from the first sutra of the Pail canon, asserting that "all phenomena in the universe are interrelated [and that] the nature of the entire universe is contained in each particle" (Keown 119–20).

Henri Lefebvre gave a foundational, even revolutionary, description of rhythm as the emergent phenomenon whenever and wherever "a place, a time, and an expenditure of energy" coincides with Einstein's concise equation for relativity ($e = mc^2$), because any analytic tool of rhythmanalysis would need, by necessity, to confront "the spectre of wavelike movements [...] from the micro to the macro, from the corpuscular movements to those of metagalaxies" (Lefebvre 90–1). Lefebvre's description of the vibrant fields around us addresses the daunting methodic challenge to any rhythmanalytic project: "Everyday life remains shot through and traversed by great cosmic and vital rhythms: day and night, the months and the seasons and still more precisely biological rhythms" (84). Those positioned at the coincidence described by Lefebvre are intertwined with/in an implicate order "particularly suitable for the understanding of unbroken wholeness in flowing movement" (Bohm 172), because they reside at the coincidence of cosmos and consciousness constructed from "vibrant matter" and poised to achieve/perceive "a more subtle awareness of the complicated web of dissonance connections between bodies" (Bennett 5).

The entanglements of form and function reflected in a seemingly straightforward song open onto rhythmic operations transferred to and operating within bodies passing through spacetime, a measurable (and memorable) passage creating ripples and generating interactions among the vibrant energies and fields through which those bodies pass and of which they are woven. Rhythm saturates and shapes the invisible and visible alike, operating across and within those thousand plateaus explored and mapped by Gilles Deleuze and Felix Guattari. Such coincidence inevitably creates entanglements where entities comprised of rhythmic processes collide, where rhythmic operations establish boundaries and conditions that thereby define the ambient fields of those plateaus through the "superposition of disparate rhythms and express from within of an interrhythmicity [without] meter or cadence" (Deleuze and Guattari 329). Without meter or cadence, connective forces working among and between discrete rhythmic phenomena create a broadened framework capable of producing "difference" in and through rhythm.

In such a guise, rhythm functions as a dynamic and transformative "vitalist force giving birth to the emergence of the new, of the event" (Crespi and Magnani 46–7) and becomes a governing dynamic that exerts force, infuses direction yet remains external to any particular manifestation of rhythm (the Deleuze and Guattari position quoted above: particular rhythms are governed by a pan-rhythm beyond specific manifestations). Generically, the concept of a boundary condition migrated to other disciplines and forms from physics, where physical boundary conditions determine the state of an environment. For example, a governing dynamic (e.g., gravity) serves as a boundary condition at the event horizon of a black hole or on an approach to the sun, because its rhythmic force captures all that comes within its pull and governs the experience of time and space in the process, conditions creating a "computational boundary of a domain." However, my approach steps beyond physics applications in hopes of giving "a fundamentally new meaning" to its applications (Paksi 6). Likewise, as implied by Grace Jones, being a slave "to the rhythm" (see prior discussion) has roots in circadian rhythms that govern the spatiotemporal unfolding of our "work, life, and love" and thereby help shape "the rhythms of [any] social experience" (Levine 51) yet remain capable of influence, as well, by "the breath of the moon" called tides (O. Jones 189). This proliferation of interwoven boundary conditions establishes the foundation and limit of what is possible, but rhythm itself (in all its guises and locales) suffuses all inner and outer events, creating conditions of being "Everything, Everywhere, All at Once" (*c.*2022).

Such circadian rhythms set boundary conditions just like gravity, although unfolding in different parts of the body and brain within (e.g., sleep cycle) and without (e.g., day and night). Of course, we break such habits at will. Nonetheless, such rhythms abide in inner and outer phenomena, establish new rhythmic practices (e.g., shift work), and alter bodily systems (e.g., digestion). Rhythm so theorized unfolds at the locale or cornerstone Alain Badiou describes as the "evental site" (*Being and Event* 176), which he argues must always be "*local* [because] every trans-formational action originates *in a point* […] inside a situation" (176), where a "unity of the multiplicity *is* the ontological identity" (Badiou, *Infinite Thought* 128). For any rhythmanalysts, the objects of study (rhythmic phenomena within and without the rhythmic researcher) engage "cyclical processes and movements, undulations, vibrations, [innumerable] returns and rotations from the microscopic to the astronomical, from molecules to galaxies, passing through the beatings of the heart, the blinking of the eyelids and breathing, the alteration of days and nights, months and seasons and so on" (Lefebvre 82, 84).

The range of Lefebvre's description of the complex obstacles for any rhythmanalysis undergoes further expansion by Deleuze and Guattari, when they turn to the transhistorical and multioperational movements of rhythm and "[its] passage from one milieu to another" (365). Rhythm distributed across and imbedded in such plateaus serve as a stabilizing boundary condition from which other discrete forms of rhythm arise, an operational force that "never [exists] on the same plane as that which has rhythm" (see above) due to its pan-system function as the fundamental boundary condition of all plateaus.

II

The "rise of rhythm" evoked in the title of this collection and within this essay has been underway for a while, with the exemplary genealogic study compiled by Paolo Crespi and Sunil Magnani confirming that "rhythm as a critical, philosophical concept has a long history," which they begin with Heraclitus and Plato and conclude with Bachelard and Lefebvre. The rise of rhythm mapped here takes a more "punctuated" path of analysis, offering what William Blake termed in his closing illuminated prophecy *Jerusalem*. "Minute particulars" (E 251: J 91.21) that shape "the outward expression in this world of the eternal individualities of all things" (Damon 250). The focus, then, is on particular *exempla* that establishes rhythm as a boundary condition, a governing dynamic that stabilizes universal operations, beginning with the extraordinary foundation laid down by the oft-overlooked Robert Fludd (1574–1637) during the Renaissance.

Robert Fludd is a fascinating figure whose scope of knowledge was on par with Leonardo da Vinci; he was "educated by private tutors" until matriculating to St. John's College in Oxford (specializing in medicine) and who, by his graduation in 1596, had already "compiled treatises on music, astrology [and astronomy]" (Godwin 3, 4). As an experimental alchemist and practicing physician, Fludd passionately believed that "the key to understanding how all things are related comes from an occult knowledge of the musical harmony throughout the universe" (Huffman 23). Fludd's explorations promoted the "expansion of the concepts of *musica mundane* and *musica humane*," which served as the foundation for his wide-ranging analyses of the entire cosmos, which was grounded in his belief that "an underlying harmony and congruity [existed] between the universe [or macrocosm] and man [or microcosm]" (23).

The sources that flowed into Fludd's system vary widely but were anchored in "his Renaissance Christian Neoplatonist" (Huffman 13) and supplemented by his blending of "Neoplatonism, Hermeticism and Christianized Kabbalah," and he believed his works provided, through harmonic interactions between image and word, the "master key" (Godwin vii) to comprehending the cosmos. As critics and scholars have noted, Fludd "took as his subject everything" (James 129), and the complexity and range of his attempt to connect all knowledge resulted in the articulation of what might be called, in current terms, a GUT (grand unified theory) or TOE (theory of everything). His fusion of such sources leads to the articulation of the universe as a rhythmic-generative vehicle stabilized by God's steady tuning of the instrument of the universe. Fludd's best-known image, "Celestial Monochord," provides the most elaborate visualization of this musical organization (Figure 3.1).

Fludd's design emphasizes the existential interconnection of "the three worlds [positioned] along the single string of a monochord," which are governed by rhythmic interactions, with the "entire cosmos thus [sounding] the two-octave gamut of Greek and medieval music theory" (Godwin 64). The progressive and projective dimensions of the design depend, of course, on the harmonizing presence of God, whose hand tunes the monochord at the head of the design. The harmony is supplemented by symmetry as well, for example his positioning of the sun at the mid-point of the design and string (the corrective offered

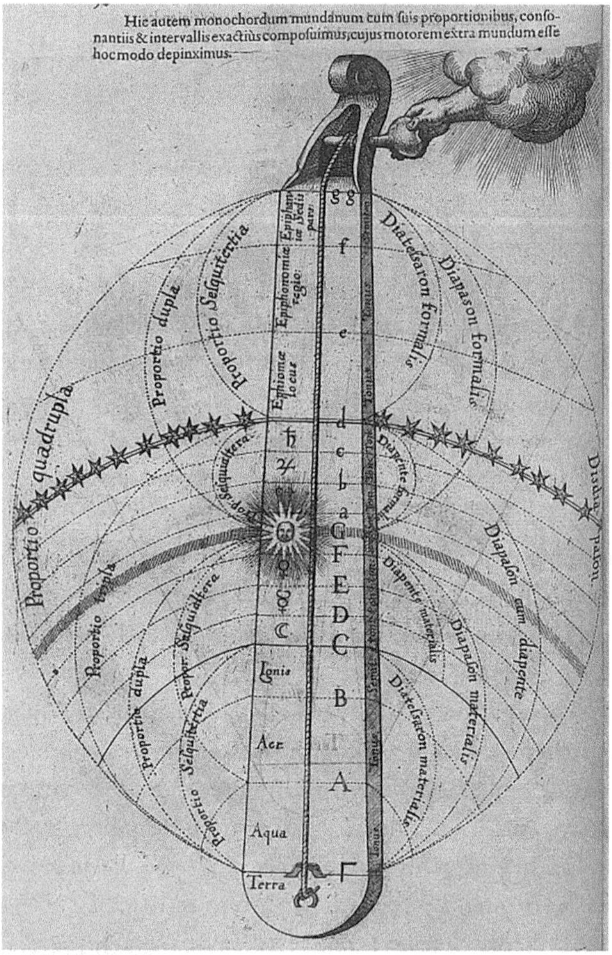

Quelle: Deutsche Fotothek

Figure 3.1 Robert Fludd, "Celestial Monochord," public domain. Image courtesy of Wikimedia Commons.

by Copernicus), which is given a human countenance (perhaps a nod to his contemporary Johannes Kepler). However, as Nobel Prize physicist Wolfgang Pauli indicates, in his comparative analysis of Fludd and Kepler, the later viewed cosmic harmony and symmetry as built upon "quantitative, mathematically demonstrable premises" and "criticized [Fludd's work] very violently [as] an archaistic-magical description of nature culminating in a mystery of transmutation" (121).

In spite of his critique of the alchemical and mystical aspects of Fludd's work, Pauli found the shared emphasis by Fludd and Kepler on music "interesting as a matter of principle" and saw in "the quaternary attitude of Fludd [vs. the trinitarian attitude of view Kepler] a psychological point of view [offering] a greater completeness of experience" (128). Pauli's further observations note that "since the discovery of the quantum of action, physics has gradually been forced to relinquish its proud claim to be able to understand, in principle, the *whole* world" (130). In Pauli's view, the "concept of complementarity" offered greater stability, since "microphysics shows [...] an uncontrollable interaction between observer and system observed" (130, 131), resulting in degrees of uncertainty.

I will later return to Pauli's description of physical operations in the discussion of Thomas Young's famous double-slit experiment, which was performed before members of the Royal Society and demonstrated that light was propagated through waves (versus particles as Newton argued). However, I turn to another earlier physician, David Hartley (1705–57), who "made the first significant foray into neurophysiology" (Finger 336) through an early mode of rhythmanalysis, by blending "Newton's vibration theory with Locke's associationism to explain memory," arguing "that vibrations in the nerves and brain [created] backward and forward motions [...] within the sensory nerves" (338). Vibration theory has a long history extending back to the origins of rhythmanalysis (Crespi and Magnani 30–1), but the rise of associationism, which was Hartley's central approach, across the eighteenth century helped generate what Alan Richardson termed "neural romanticism" (4–5). Outer vibrations or stimuli, for Hartley, become inner "vibratiuncles" that are impressed onto the medullary material of the brain, thereby promoting a purely mechanical/medical explanation for memory.

Medical practitioners were, thus, at the center of the "concept of rhythm emerging between 1760 and 1839 [and who viewed it] as a relationship in time" (Wellman 323). Rhythm also grew in complexity as a process defined as the "tension between wholeness and interruption" (relative to its essential duality), its startling temporal bidirectionality, and its feature of regularity (versus uniformity). Therefore, "[rhythm] was a completely autonomous movement [and] variable at every moment of its interrupted sequence," which through these dynamic elements "added a new dimension to the biological world" as these were known at the time (323, 324). Medicine serves as the pre-eminent vehicle in the emergence of a "rhythmic episteme [...] anchored in human nature" (20, 21), and Richardson provides an abbreviated genealogy of physicians (within the time frame Wellman explores) who "helped transform the psychological thinking of the time," an august group including F. J. Gall, Erasmus Darwin, Charles Bell, J. G. Spur Heim, David Hartley, Julien Affray de La Mutrie, and J. G. von Herder, who were "all [...] medical doctors [and] all were committed to the biological account of mind" (5). For Hartley, the material processes of perception were inscribed physiologically (as vibrations) that proceed from the site of sensation along the network of nerves in passing to the brain, where such vibrations meet and reinforce or alter prior oscillatory information. One can feel the hardened sense of a boundary condition (i.e., a physical law like gravity) begin to break down, as rhythmic operations emerge as both hardware and software in the human condition.

III

As a transition, let me reengage Pauli's evocation of complementarity (above) and its grounding in rhythmic operations, as he seems to sense that the shared interest in music by both Fludd and Kepler held implications—not just because of the exclusion principle that bears his name but an anticipation of his collaborative study of synchronicity with Carl Jung. Although both Jung and Pauli supported the idea that while "events in general [were] related to one another [...] causal chains," both such events periodically manifest other "kind[s] of meaning cross-connections" and began to theorize a form "to a psychic relativity" (Jung 11, 17). The idea that the music in the brain and the song of the spheres were interconnects moved Pauli, whose work on quantum to open the twentieth century unfolded along the rhythmic lines of thoughts that motivated both Fludd and Kepler.

In many ways, the final piece of knowledge that confirms the explanatory power of rhythmic operations points directly to one figure, the polymath Thomas Young, whose experiments on light at the beginning of the eighteenth century confirmed its wave aspect, thereby overthrowing the standard particulate Newtonian view with light as a wave and preparing science for the advent of the quantum age. In a January 16, 1800, letter to Edward W. Gray (then Secretary of the Royal Society of London), Thomas Young proposed a series of lectures titled "Outlines of Experiments and Inquiries respecting Sound and Light" (106). The lectures would begin with the presentation of "a few observations of the subject of sound [...] and to make as many experiments [...] as circumstances enable me to do" (106), including investigations into "direction and velocity" of light and sound, "the nature of sound," the "analogy between light and sound," "the frequency of vibrations constituting a given note," and "the vibrations of rods and plates" (106–7).

Young's experimental and intellectual range was, as his biographer Andrew Robinson suggests, daunting and dictated his title: *The Last Man Who Knew Everything: Thomas Young, the Anonymous Genius Who Proved Newton Wrong and Deciphered the Rosetta Stone, Among Other Surprising Feats*:

> Not only did he make pioneering contributions to physics (the wave theory of light) and engineering (the modulus of elasticity), to physiology (the mechanics of vision) and to Egyptology (the deciphering of the hieroglyphs), but was a distinguished physician, a major scholar of Greek, a phenomenal linguist, and an authoritative writer on all manner of subjects, from carpentry and music to life insurance and the ocean tides. (Robinson viii)

Reading each aspect of his abilities, one cannot overlook its engagements with a wide range of rhythmic phenomena—wave theory and elasticity to music and ocean tides. Each has a rhythmic element: the circadian play of the tides, the duality of light, the harmonics and notes of music, the rhythmic differences within languages, or the relentless periodicity of actuary tables. That said, Young acknowledged in an 1802 letter to a friend (Abdrew Dalzell) that "the theory of light and colors, though it did not occupy a large portion of time, I conceive to be of more importance than all that I have every done" (Robinson 95). At that moment, his experimental outcomes dethroned the dominance of Newtonian dependence on a particulate view of light and, in the process, prepared the path to contemporary physics, especially quantum theory, through the demonstration of patterns of interference as a byproduct of his famous double-slit experiment.

At the outset of the nineteenth century, there remained few proponents for the wave theory of light, although Christian Huygen's, 1678 paper remained read and relevant, a position already implied in the surprisingly sophisticated thought of Robert Fludd, eighteenth-century thought, whether in the medical opinions of David Harley or Thomas Young's analysis of the wavelike aspects of light (Figure 3.2).

The experimental outcomes—wave–particle duality, the presence of both constructive and destructive patterns of interference—produce patterns of interference; however, when one tries to establish which hole a particle/wave of light has passed through, the pattern disappears, thereby confirming that the act of observation itself is entangled with the outcome, an outcome that lurks at the heart of what George Gamow termed *Thirty Years that Shook Physics: The Story of Quantum Theory* (c.1900–1930). Although this long-term outcome inspired some of the best minds in physical theory (e.g., Albert Einstein, Max Plank,

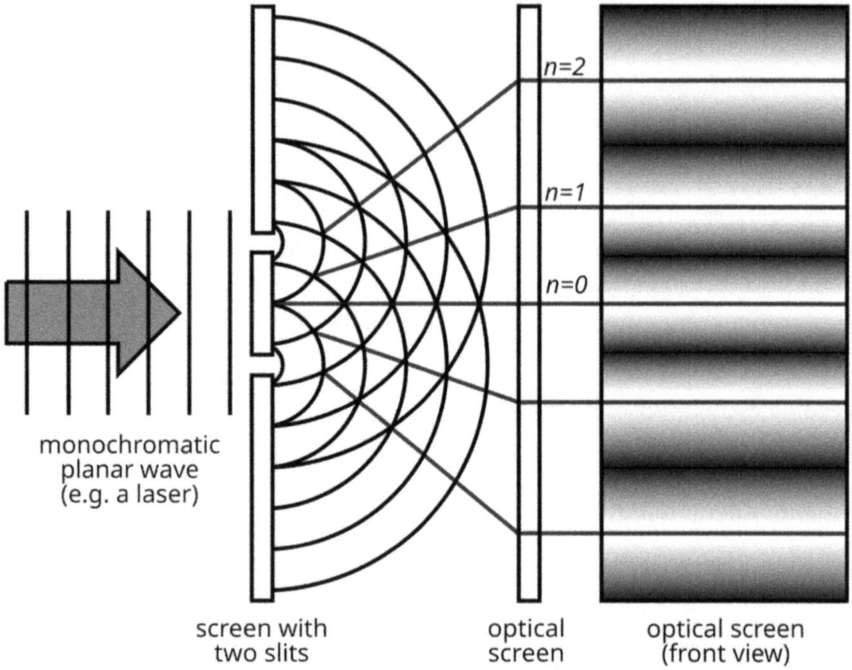

Figure 3.2 "Double-Slit Experiment," public domain.

Niels Bohr, Louis de Broglie, Erwin Schrödinger, and Werner Heisenberg), the immediate effect was an imaginative liberation from Newtonianism, although Young paid dearly for this "affront" by being denounced by Henry Brougham (the future prime minister) in the pages of the *Edinburgh Review* (which he partially owned). However, what followed immediately was a veritable explosion of experimental outcomes that both used rhythmic methods and discovered rhythm lurking in almost every discovery prior to 1905 (Einstein's famous year).

IV

The presence of rhythmic phenomena within varied layers of cultural production in the last half of the eighteenth century certainly confirms Janina Wellman's brilliant observation—based on her analysis of the developmental emergence of embryology during the last half of the eighteenth century—that the world entered "a rhythmic episteme" (20) across the eighteenth and nineteenth centuries and that remains a constant which endures in our own moment. My argument for rhythmic operations as the fundamental component or boundary condition for encoding what William Hazlitt termed "the spirit of the age" returns (inevitably I would say) to Henri Lefebvre's famous characterization of rhythm (see above) as an "interaction between a place, a time, and an expenditure of energy" (15). Notably, Lefebvre's adoption and adaptation of the formal elements of Einstein's best-known equation for the theory of relativity ($e = mc^2$) should not go unnoticed because its intersecting elements provide a method capable of stimulating deeper and wider attention to oscillatory and vibratory phenomena within and without.

Although somewhat primitive and grounded in erroneous assumptions connected to Newtonian theories of the ether, Hartley's expansive and exploratory model gained the praise of Joseph Priestley, who declared that the work opened onto a new world of interiority to articulate "the science of man." Although the work was fraught with flaws, its core elements (vibration and association) had great explanatory power that did indeed break into a new world where external and internal dynamics proved mutually conditioning and illuminating. As well, Hartley's fundamental insight to adopt the interactive model of rhythmic operations without and within remains in harmony with current models as evidenced by Evie Bentley's more recent description that "we live in a rhythmic world, [. . .] whether or not we recognize [such rhythmic operations]" (11).

Fifty years later, in a series of ground-breaking experiments and lectures delivered to the Royal Society (*c.* 1799–1802), Thomas Young performed perhaps the best-known experiment in the history of physics, the double-slit experiment. Young's experiment, which projected light through two slits onto a screen, yielded an outcome that challenged the dominance of Newton's particulate theory of light, which demonstrated that light acted under different conditions both as a particle and a wave, and which exhibited the prominence of patterns of interference. When Young published his findings in *Philosophical Transactions of the Royal Society*, the results drew the ire of the political heavyweight and champion of Newtonianism, the future Lord Chancellor Henry Brougham, who excoriated Young in the pages of the *Edinburgh Review* (the journal he helped found in 1802).

However, in spite of such vitriol, Young's model eventually took hold, with the duality of light as both a particle and a wave (better known as complementarity) providing a throughline of thought to the work of Werner

Heisenberg and Niels Bohr at the beginning of the twentieth century. Young's outcomes had, through the distinctive discovery of patterns of interference, proved a significant tool for future cosmology, because the distinctive pattern of interference led to the phenomena of red shift, which allowed Georges Lemaitre to hypothesize the current big bang cosmological model (in 1927) and became thereafter the essential tool used by Edwin Hubble to propose the expanding universe, which remains our standard model. In the immediate wake of Young's copious contributions to science and as the assistant to the renowned chemist Sir Humphrey Davy, the self-taught and ever-humble Michael Faraday began experimental studies that explored rhythmic properties of electrolysis and that extended to the discovery of electromagnetic fields. Relative to the work of Young, Faraday also contributed numerous breakthroughs in chemistry and was the first to discern that magnetic fields could actually exert an overt impact on light, thereby establishing an inherent relationship between two seemingly different aspect of oscillatory and vibrational phenomena, setting the stage for the work of James Clerk Maxwell.

Maxwell took the work of Faraday and offered refinements that literally propelled physical theory forward through dynamic models grounded in material rhythmic operations. Maxwell was, for example, the first to comprehend that electricity, magnetism, and light were essentially different materializations of the ubiquitous rhythmic forces at work within the deep structure of observable events. In 1865, with the publication of *A Dynamical Theory of the Electromagnetic Field*, Maxwell extended his initial insights on the wave aspects of observable phenomena by demonstrating that electrical and magnetic fields were propagated through space as waves moving at the speed of light. Maxwell's work formalized rhythmic dynamics, thereby paving the way for the rapid, sequential emergence of all aspects of quantum mechanics, including quantum chemistry, quantum field theory, quantum technologies, and quantum informatics. Thus, although the opening decades of the twentieth century are often labelled the "thirty years that shook physics" (in homage to the excellent book of the same title written by Soviet and American polymath, theoretical physicist and cosmologist, George Gamow), the fifty-year span from Young's double-split experiment to Maxwell's unification of electricity and magnetism could equally be called "the fifty years that shook physics."

This type of energy can certainly be felt in the poetry that inaugurates the nineteenth century and that reflects this spirit of the age, although it does not necessarily reflected William Hazlitt's best-known book of the same title, which seems to miss what might be termed (again referencing the exemplary work of Janina Wellman) the pulsations and rhythmic manifestations of the multitudinous "forms of becoming" (28–9) expressed "[as] loud and whirlwind harmony" that transmits this spirit "through empty space, music and light" (P. Shelley 27: 237–40). Shelley's description of celestial mechanics (large and small/ without and within) in the closing act of *Prometheus Unbound*—those rhythmic harmonies propagated through space and time within material forms (music and light) that are "entangled" with our bodies and minds (in Karen Barad's sense of meeting the universe halfway [247])—reflects his deep and wide knowledge of the contemporary sciences discussed above, as well as his participation in an alternative romantic revolution in poetic language grounded in science. Erasmus Darwin announced his commitment to articulate a lyricized science, and Shelley follows this impulse, while leaving behind its most antiquated formal elements within Darwin work while plundering his footnotes, which most interested Shelley.

Shelley's aesthetic weaving of such elements provided a foundation for an imaginative leap toward a visionary physics, a term most often associated with William Blake's mini-epic poem *Milton*, in which through the agency of the imagination he strives to "Build the Universe stupendous: Mental forms creating" (30:20). Like Shelley, Blake knew Darwin's mode of lyricized science quite well and even completed designs for Canto III of *The Economy of Vegetation*, and like Blake, Einstein often asserted that "imagination is more important than knowledge," with this short proclamation rippling beyond scientific environments and marketed through T-shirts, bookmarkers, mouse pads, and posters. Althouh this might seem strange, we need only recall his daily use of imagination, whether visualizing what it would be like to sit astride a beam of light while taking the train to his day job at the patent office in Bern or engaging through his famous thought experiments with Niels Bohr during their intense arguments over quantum physics. Strangely, although this evocation might not seem scientific, Blake actually remains the most cited poet in books, essays, and internet posts published by scientists of every discipline, with special fondness for the opening quatrain of "Auguries of Innocence":

> To see a World in a Grain of Sand
> And Heaven in a Wildflower
> Hold Infinity in the palm of your hand
> And Eternity in an hour

By my last count (up to last year), Blake's quatrain has appeared in over 100 scientific works during the last seventy years to explain a wide range of physical phenomena, and when I once visited the Palomar Observatory (*c.*1996) it was posted on four doors.

The rhythmic view threaded throughout this essay, as its migrates from mystical speculative in the Renaissance through scientific dethroning of Newtonian physics through the experimental to poetical articulations sciences during the long nineteenth century, continues to shape the contours of the real as a boundary condition, one finally recognized and understood as deeply interconnected and manifest in interactivity often bundled with temporal and spatial energies. This expresses a unified field defined through relativistic lines of flight (of thought) entangled across disciplines and discourses, and the detection of diverse forms of rhythm across these as they interact top to bottom, side to side, within and without, a flow slowly flowing from an emergent episteme described by Wellman.

These rhythmic properties also provide focus for the transdimensional propagation described above by Deleuze and Guattari, who connect particular "critical moments" traceable within varied expressions of rhythm. Although inevitably incomplete, rhythm thus propagated "from one milieu to another" (313) articulates "both a human universal and [a] cultural construct" (Barletta 161–62) and connects ways of being, as when Georg Sulzer "situated rhythm in man's *physis* [...] the physiological and psychological constitution in equal measure, physical and intellectual motion, man's existence as a feeling being whose perceptions spring from his sensory physiological disposition" (Wellmann 71) yet can, as a result, prove "punishing" and "pervasive" (Levine 49). As reflected across this collection, rhythm is "a category that aways already refuses the distinction between aesthetic form and other forms of lived experience" (53), which returns to the implications for a "simple" song like *Slave to the Rhythm*. Lefebvre imagined that future "rhythmanalysts" would listen "first to the body," the primary zone of contact with alterity, with the body–mind interface serving "as a metronome" (29).

The pulsating and oscillatory forms of rhythmic phenomena across all its understood platforms (within and without, near and far) offer significant scope for a method to map the interacting spectra of the real, although the Lacanian "real" should never be confused with "any normative reality principle" and "is not of this world," as "topology could never be an ontology" (Thurston 143). As the epigrams suggest, rhythmic outcomes emanate from every event (especially as understood through Alain Badiou) and involve interactions of consciousness, energy, space, and time, a view in accord with most current models of consciousness and the cosmos. Inevitably, then, such a method must position itself along the event horizons where these elements collide and cohere, where dynamic interactions and patterns of interference produce complex patterns open to rhythmic analysis and lead into and out of the interwoven plateaus of real, symbolic, and imaginary processes as these categories passed from Jacques Lacan to their seemingly endless extension across the last two generations of critical thought.

Works Cited

Aviram, Amittai F. *Telling Rhythm: Body and Meaning in Poetry*. U of Michigan P, 1994.

Badiou, Alain. *Being and Event*. Translated by Oliver Feltham. Continuum, 2007.

Badiou, Alain. *Infinite Thought: Truth and the Return of Philosophy*. Edited and translated by Oliver Feltham and Justin Clemens. Continuum, 2005.

Barad, Karen. *Meeting the Universe Halfway: Quantum Physics and the Entanglement of Matter and Meaning*. Duke UP, 2007.

Barletta, Vincent. *Rhythm: Form and Dispossession*. U of Chicago P, 2020.

Bennett, Jane. *Vibrant Matter: A Political Ecology of Things*. Duke UP, 2010.

Bentley, Evie. *Awareness: Biorhythms, Sleep, and Dreaming*. Routledge, 2000.

Blake, William. *The Complete Poetry & Prose of William Blake*. Edited by David V. Erdman. Anchor Books/Doubleday, 1988.

Bohm, David. *Wholeness and the Implicate Order*. Ark, 1983.

Crespi, Paola, and Sunil Magnani. "A Genealogy of Rhythm." *Rhythm and Critique: Technics, Modalities, Practices*. Edited by Paolo Crespi and Sunil Mangini. Edinburgh UP, 2020.

Damon, S. Foster. *A Blake Dictionary: The Ideas and Symbols of William Blake*. Shambhala, 1979.

Deleuze, Gilles, and Felix Guattari. *A Thousand Plateaus: Capitalism and Schizophrenia*. Translated by Brian Massumi. U of Minnesota P, 1987.

Edensor, Tim. "Introduction: Thinking about Rhythm and Space." *Geographies of Rhythm: Nature, Place, Mobilities and Bodies*. Routledge, 2016.

Finger, Stanley. *Origins of Neuroscience: A History of Explorations into Brain Functions*. Oxford UP, 1994.

Fludd, Robert. *Essential Readings*. Edited by William H. Huffman. The Aquarian P, 1992.

Godwin, Joscelyn. *The Greater and Lesser Worlds of Robert Fludd: Macrocosm, Microcosm & Medicine*. Inner Traditions, 2018.

Huffman H. William, Ed. *Robert Fludd: Essential Readings*. Aquarian/Thorsons, 1992.
James, Jamie. *The Music of the Spheres: Music, Science and The Natural Order of the Universe*. Copernicus/Springer-Verlag, 1995.
Jones, Grace. *Slave to the Rhythm*. Island Records, 1985.
Jones, Owain. "'The Breath of the Moon': The Rhythmic and Affective Time-spaces of UK Tides." *Geographies of Rhythm: Nature, Place, Mobilities and Bodies*. Edited by Tim Edensor. Routledge, 2010.
Jung, C. C. *Synchronicity: An Acausal Connecting Principle*. Translated by R. F. C. Hill. Princeton UP, 1973.
Keown, Damien. *Dictionary of Buddhism*. Oxford UP, 2003.
Krauss, Nina. *Of Sound Mind: How Our Brain Constructs a Meaningful Sonic World*. MIT P, 2021.
Lefebvre, Henri. *Rhythmanalysis: Space, Time and Everyday Life*. Translated by Stuart Elden and Gerald Moore. Bloomsbury Academic, 2021.
Levine, Caroline. *Forms: Whole, Rhythm, Hierarchy, Network*. Princeton UP, 2015.
Paksi, Dániel. "The Concept of Boundary Condition." *Polanyiana*, vols. 1–2, no. 23, 2014, pp. 5–20.
Pauli, Wolfgang. "The Influence of Archetypal Ideas on the Scientific Theories of Kepler." *Robert Fludd: Essential Readings*. Edited by William H. Huffman. The Aquarian P/Harper Collins, 1992.
Patel, Aniruddh D. *Music, Language and the Brain*. Oxford UP, 2008.
Richardson, Alan. *British Romanticism and the Science of the Mind*. Cambridge UP, 2001.
Robinson, Andrew. *The Last Man Who Knew Everything: Thomas Young, The Anonymous Genuis Who Proved Newton Wrong and Deciphered the Rosetta Stone, Among Other Surprising Feats*. Plume/Penguin, 2007.
Shelley, Percy. *Shelley's Poetry and Prose*. 2nd ed. Edited by Donald Reiman and Neil Fraistat. W. W. Norton & Company, 2002.
Thurston, Luke. "Ineluctable Modalities: On the Borromean Knot." *Key Concepts of Lacanian Psychoanalysis*. Edited by Dany Nobus. Rebus P, 1997, pp. 139–63.
Wellman, Janina. *The Form of Becoming: Embryology and the Epistemology of Rhythm, 1760–1830*. Translated by Kate Sturge. Zone Books, 2017.
Young, Thomas. "Outlines of Experiments and Inquiries Respecting Sound and Light." *Philosophical Transactions of the Royal Society*, vol. 90, 1802, pp. 106–50.

4

Textural Rhythm and Textural Sense-Making

Sha Xin Wei

Approach and Overview

Rhythm's structuring of the temporal aspect of experience invites the question: what sorts of sense are enabled by rhythm? This chapter offers a progressive relaxation from various geometric tropes of time, such as the conceit of periodicity, with the aim of developing a pre-individuated, field-like notion of the variation of dynamical matter, energy, and affect, a textural rather than geometrical rhythm. Unmooring from hylomorphic and thus quintessentially reductive approaches to structuring experience evokes a very different mode of subjectivity: a textural subjectivity that contingently constructs lived, durational occasions out of boundlessly multiple textures of matter, energy, sense, and value, ever-varying not with respect to some abstracting dimension, but with respect to each other.

I approach this in William James's radically empirical spirit, propositionally employing Alfred North Whitehead's ontological principle: build concepts not as transcendent abstractions but as instances of concrete experienced occasions. An idealist approach in the genealogy of Western metaphysics would seek to explain the world and experience in terms of *a priori* categories such as subject and object, starting with transcendental forms and principles and explaining particulars in terms of those abstractions. However, robustly avoiding *a priori* ideal forms would also include avoiding a notion of subject, and object, prior to an occasion of experience.[1] This problematizes the question of *meaning* by shifting attention to the emergence of *sense* in the ever-developing magma of experience without recourse to *a priori* subject, mind, consciousness, agent,

1 See James, "Does 'Consciousness' Exist?".

or other primitives of logical positivist and cognitivist theory obscuring that which they purport to explain. Technologies designed under idealist conceits, such as genetic engineering or gamified sociality, would impose *a priori* ideal forms on inert matter, the essential reductive move of *hylomorphism*. By contrast, Gilbert Simondon's approach to technical ensembles, individuation, and ontogenesis takes a thoroughly *processualist* perspective: instead of asking what the world is made of and what categories of things there are in the world, we ask how things come to be, and think in terms of transformation, dynamics, processes. This processualist attitude provides traction on the most essential aspect of the experienced world: its temporality.[2] It is under this perspective that the study of *rhythm* takes on deeper implications. To be thoroughly empiricist about experience as dynamical, developmental, processual also motivates an experimental approach to studying phenomena such as the dynamical development of sense in live, experienced occasions. And it is more consistent for me and my research-creation collaborators to construct and experience these occasions for ourselves as well as for others, and then publish considerations based on both reflection and experience. For example, I build on work by Garrett Laroy Johnson on rhythm and textural temporality (Sha and Johnson). Al Bregman, a pioneering researcher in auditory illusions, observed that in psychoperception studies, there is no substitute for the researcher to experience themselves phenomena that are intrinsically *subjective*.[3] (In place of using "subjective" as an adjective, I prefer to say the phenomena co-articulate the subject.) In my view, what Bregman said of phenomenological science goes for nonreductively experimental philosophy as well. That's why this essay is equally informed by twenty-five years of experiential experimental installation-occasions created by artists in the Topological Media Lab (TML) and the Synthesis research-creation ateliers.[4] Why experimental? A singular work of art or performance can powerfully evoke sense, but my institutional intention building the TML and Synthesis has been to develop methods for a *scienza*

[2] This processualist attitude draws on a body of thought that includes Zhuangzi, Laozi, Heraclitus, Lucretius, Karl Marx, Henri Bergson, Alfred North Whitehead, Gilbert Simondon, and Gilles Deleuze, and contemporary philosophers Isabelle Stengers, Elizabeth Grosz, Pascal Michon, Thomas Nail, and David Morris.

[3] Bregman, "ASA at McGill, CIRMMT Keynote," 2008.

[4] See topologicalmedialab.net and synthesiscenter.net. Research-creation is a term of policy from Quebec and then adopted at the federal level in Canada referring to institutionally organized production and dissemination of portable knowledge by artists, performers and creative writers. By portable I mean the knowledge can enable researchers in other domains of inquiry to generate more knowledge (FQRSC; Chapman and Sawchuk; Sha, "What Is Research in Contemporary Art").

nuova with which to richly study phenomena of experience where we can reproduce not an experience—an impossible aspiration in a dynamical world—but the *conditions* of experience. This is why we assembled and as necessary invented techniques for modulating the felt experience of an occasion. And because we have been interested in the temporality of experience, we have used multiple modalities of *time-based media*: visual imagery, sound fields, theater-grade lighting, acoustics, tangible props that *dynamically vary in concert with the activity of the inhabitants* of the space that is saturated with those media. (We call such media *responsive media*.) Throughout this essay I will motivate and illustrate what I will call *textural sense-making* by referring to experimental experiences/experiential experiments, many of which are installation-occasions explicitly created by the TML and Synthesis to enable participant artist-researchers to invent, discover, and experience the textural phenomena this essay addresses. We needed to invent novel multimodal techniques to step aside from conventions of spectatorship, screenism, linguistics, literary form, image and film, which would return the experiment's inhabitants to canonical formal interpretations of the experimental experience. In other words, where Bertolt Brecht employed an *Verfremdungseffekt* at the level of roles, symbols, and apparatuses, my topological media approach enables a parallel defamiliarization at the level of intensity and a signifying semiotic (to borrow notions from Deleuze and Guattari). I have elaborated these experimental methods elsewhere (The Atelier-Lab; *Poiesis*). Here I sketch the experience of a handful of installation-occasions to introduce phenomena *in situ* as they relate to the conceptual arc of this essay.

To augment the processualist and radical empiricist conceptual approach, I introduce an exact dash of *non-discrete topology*, and construct a field-based, *textural* approach to rhythm as a mode of sense-making. This textural concept of rhythm provides a grip on ontogenetic structuration, in which matter *is* the structuring of matter, en route to a textural account of *sense-making* inspired by Noah Moss Brender. For a more adequately robust notion of sense, I make use of David Morris's developmental ontology, and Deleuze's concept of sense.

From Entities to Textures

Let me describe some experimental experiences that foreground the textural aspect of experience.

Experience 1: *Becoming Cloud*

The Synthesis atelier created the *Becoming Cloud* installation[5] (Figure 4.1) in a large theater-scale black box space. You stand in front of a scrim, a sheer cloth hanging from the ceiling of a black box, spanning 20 feet (7 meters) across a large space. As milky clouds coil across the scrim you hear hisses building left and right punctuated by occasional crackling in the sonic field, an atmospheric sonic and visual landscape.

As you walk to one side, the scrim darkens with mist, denser where the shadow of your figure is moving more, and sparser where less. You begin to appear in silhouette. This dark mist lifts as if it were a mass of hot moist air, and as it rises it cools and condenses into white cloud. The swirls of cloud becoming visible also stir up banks of sound that mass according to the local and momentary dynamics of the mist. These banks of sound emerge spatially so that you are enveloped in an atmosphere of sound; the dynamics of the cloud to your left and right drive the dynamics of the sonic field across the width of your embodied space. As other people walk to your side their movement billows the atmosphere. As you settle into motionlessness, the mist begins to dissipate, but the atmosphere's dynamic persists as do the sound's spatial dynamics.

Figure 4.1 "Becoming Cloud." Image by the author. See https://vimeo.com/synthesiscenter/clouds.

[5] Synthesis, *Becoming Cloud* installation-occasion, https://vimeo.com/synthesiscenter/clouds.

As a responsive installation, *Becoming Cloud* is not an illustration or analogy of a concept but an occasion for enactive, embodied *texturing* primordial to linguistic mark-making and meaning-making. After experiencing this installation, one can ask with more sensitivity: How does sense appear before and after a gesture? How does this occasion make sense before and after a body?

From Agent to Texture

According to an interactionist account of *Becoming Cloud*, you are "drawing"—making a mark in the mist—but there's an array of computations ranging from feature extraction based on sensor data (e.g., camera-based computer vision) to the highly nontrivial weather simulation (Bryan; Mechtley et al.) that constitute a deterministic system—deterministic because it is algorithmic with no pseudorandom component—but with unpredictable consequence. All these processes run *concurrently* in concert with the movement of the visitors' bodies, and so it is fruitful to take a more ample and "symmetrical" interpretation of the dynamical situation. This interpretation offers a nonanthropocentric, nonreductive approach to dynamic and sense-making that avoids appealing to ideal forms from psychology, language, and cognitivism.

Experience 2

First, let's perform a parable cartooning the graph-theoretic agent-based interactionist accounts of social and computational dynamics. Ink some dots representing social-psychological subjects onto a piece of tissue paper. Draw arcs joining the dots to represent social relations. Pick one dot to represent yourself and ask, who are the others? The other dots. Others are selves like myself. Now delete that one selected dot representing yourself and ask, who's the complement to the self? It is not just the other dots—selves like you, not just the arcs relating to others like you, but the entire sheet of tissue, minus that one deleted point representing yourself. The complement to myself is more than other entities like myself, more than the suite of relations among other selves, but also the material, social, affective *plenum* in which all these entities emerge or dissolve. This parable that you can perform for yourself illustrates how the proper complement to self is not other selves but the world in plenist density.

This suggests that we be alert to anthropocentrism when we try to understand phenomena of experience.[6]

From Event to Texture

As we problematize the concept of "agent" in this way, we will also problematize "thing" as event. In *Difference and Repetition*, Deleuze writes:

> The difference "between" two things is only empirical, and the corresponding determinations are only extrinsic. … Lightning, for example, distinguishes itself from the black sky but must also trail it behind, as though it were distinguishing itself from that which does not distinguish itself from … [the lightning].

So once again, we are led back to consider not the conventionally positivist notion of event as foreground, but the generativity of the "black" from which the event distinguishes itself, a black which is not void, and does not distinguish itself from any thing. (The space between "any" and "thing" is significant.)

If we recognize things, to adopt Whitehead's phrasing, as drops of experience, we can think of things as events, in which case there is a deep temporality to things in their processuality. This hints that, understood in a suitably robust way, rhythm may constitute objects of experience. In order to develop that proposition, I relinquish the idealist notion of metrically regular period—"beat"—in favor of a more primordial approach to recurrence, which depends in turn on the phenomenon of difference.

In its most naive conception, rhythm could be said to be the repetition of some event, like a palm striking another palm, or the return of a musical tonality to a key, or the recurrence of the constriction of the heart—a beat. But then how do we recognize one event as the same as another so that it could be said to be a repetition of the other? What is the rhythm of a cosmos ensuing from a configuration to which the cosmos never returns? This suggests that a deeper consideration of rhythm would attend to the night from which an event distinguishes itself. We will return to this when we take up Deleuze's notion of sense later.

As a start, let me propose rhythm as the relative motion of a body across spatial–temporally felt inhomogeneity. Without inhomogeneity or movement there is no

[6] Of course, there's no need to deny that I speak as a human being, and therefore my thoughts are anthropomorphic. But they need not be anthropocentric. Mathematics, Whitehead's process philosophy, and Isabelle Stengers's cosmopolitics are examples of nonanthropocentric thought.

rhythm, and without a body, there is no sense of rhythm. Indeed, regularity or periodicity is formally convenient for signal processing and canonical forms in poetic text, but by no means that which animates song, speech, or other modes of living articulation. It was not until Plato that ῥυθμός/"rhuthmós, which was previously considered as an ephemeral disposition of something varying, an 'improvised, temporary, changeable form,' or a 'way of flowing,' became 'an ordered sequence of movements' subject to 'numbering' and 'divided into alternate times'" (Michon 4).[7] After Deleuze and Guattari we can also reinterpret rhythm in terms of variations of intensity in strata of consistency.

Analog Time: Beyond Beat

Starting with the conventional conception of rhythm as a regularly periodic, unidimensional pattern, I will progressively enrich it to be more adequate to the textural dynamics at play in an occasion of experience. Our first step is to relax the conceit that rhythm is composed of regular periodicities.

In the late sixteenth century, the story goes, Galileo Galilei noticed that the longer chandeliers suspended in Il Duomo cathedral of Pisa seemed to take longer to swing back and forth than their shorter cousins. He timed their period according to his own pulse. I focus here not on his discovery of a formula relating the length of a pendulum to its period, but on his comparing duration to his own heart's activity. Before the mechanization of "time-keeping," durational experience was ever a matter of comparing one material analog process with another. Indeed, the term "time-keeping" is itself a telltale of the process of idealization synchronized with the externalization of duration into mechanical rather than living process. What's interesting here is that the invention of the mathematical formula prefiguring centuries of idealization of movement and duration into what we call rhythm and time was itself calibrated first by the movement of blood driven by a living heart, an organ whose movement exquisitely paces the combined effects of intention, expectation, emotion, metabolism, physique, and action in the world.

[7] Pascal Michon cites Émile Benveniste's etymological point that in the seventh to fourth centuries BCE, the stem—(θ)μός "'does not designate the fulfillment of the notion but the particular modality of its fulfillment.' That is the reason why it is 'appropriate for the *pattern* of a fluid element' and commonly denotes an 'improvised, temporary, changeable form.'" This more subtle point accommodates an account that attends to the manner as well as the product of a process.

Experience 3: Einstein's Dream

During the TML's *Einstein's Dream* workshop[8] on altering the inhabitants' sense of passage, pace, and duration, at one point the video programmer asked if we might cue the audiovisual media by the computer's clock. However, instead of indexing synthesized sound and imagery on computer time, we calibrated the media according to what I called *analog time*. For instance, when the performer lit the candle (Figure 4.2), the continuous flaring of its flame modulated the "wind" that was synthesized in "real-time." As the candle slowly extinguished, the sound of wind turned into a long inhalation and exhalation that eased according to the contingent accumulated activity of the flame.[9] So the pulse and variation of the sound emerged as a poetic blend of multiple layers of contingent analog activity, quasi-physics, and live intentional action intertwining contingent bodies with tangible theatrical props. When the performer slowed her gesture the synthesized media began to subside. As she moved more animatedly, the textures of sound and shadow intensified and rippled just as water would. However, the composers artfully crafted latent responsivities of the time-based media that water could never exhibit.

Figure 4.2 "Candle-lighting installation/performance," *Einstein's Dream*, rights owned by author. See https://vimeo.com/tml/candle.

[8] See https://vimeo.com/tml/einsteindreams.

[9] See *Candle-Lighting, Einstein's Dream*, https://vimeo.com/tml/candle.

Beyond Unidimensionality

Next, we detach a second formal conceit, that of the unidimensionality of temporality.

Experience 4: 100 Metronomes

Ligeti's *Poème Symphonique* begins as an indeterminate wash of noise that emerges out of the massed sound of 100 analog metronomes, each winding down in its own way according to its accidents of matter and mechanism, accidents of how it is wound up and triggered by human performers. Most importantly, each listener assembles the felt sense of rhythm(s) as an amalgam of contingent physical, physiological, and phenomenological process.[10] It makes no sense to write down the 100 metronomes' activity as if they were 100 distinct voices, because the handful of discernible rhythmic "lines" do not correspond to the metronomes but constantly shift from moment to moment, and differ for each listener in each occasion. Rather than conceiving an abstract scheme of relations among preexisting atomic entities such as beats or time signatures, sound artists since the twentieth century have expressively worked with *texture*—ever-changing gradients and distributions of timbre, intensity, and noise. There is no score to a predetermined set of musical voices and events, thus no predetermined set of variations in sonic amplitude that could be electromechanically measured by some device. Indeed, the physical energetic resultant is a chaotic function of the analog mechanics that the human hears as noise for the first part of this performance-without-a-performer, the expressive counterpoint to radical empiricism's experience-without-a-subject. The rhythms discerned are generated *but not determined* by the mechanics; they also depend on the metabolism (a biological contingency) and the expectation (a social and phenomenological contingency of the listener).[11] Therefore, the question of how

[10] Ligeti, *Poème Symphonique for 100 Metronomes*, www.youtube.com/watch?v=NBF7QWmSOKo, Frac Franche-Comté, Conservatoire à Rayonnement Régional du Grand Besançon Métropole, April 12, 2014.

[11] Furthermore, a listening individual is not what a theorist in phenomenology or sound studies may call a hearing subject. More subtly, as Cecile Malaspina interpreted Simondon's distinction between subject and individual: "It is also true that the subject (capable of being a cause) is reducible to the notion of the individual neither for Simondon or Spinoza. For Spinoza the subject is *natura naturans*, for Simondon, likewise, the subject is 'vaster than the individual, richer than it, and having, in addition to the individuality of the individuated being, a certain weight of nature, of non-individuated being'" (14).

many degrees of freedom—that is, the dimension—the rhythm has is impossible to answer. *Poème Symphonique*'s rhythm, as heard in each live occasion, is not unidimensional; it is not infinite dimensional; it is strictly indeterminate.

Humberto Maturana and Francisco Varela made a different and more pointed argument in an appendix to *Autopoiesis and Cognition* about the artifactuality of time as a sequential dimension. Their phrasing and logic being quite precise, I'll stay close to their formulation. First, they observe that in an organism's mass of concurrent metabolic processes, a "behavioral distinction" invents a "referential dimension" that is sequential. The behavior to which they refer is a change of relation between an observer (which could be the organism itself) and the organism in its interaction with its ambient. Crucially, this sequence, which is an artifact of the distinction between "otherwise equivalent interactions," is "not in the operation of the nervous system as a closed neuronal, network" but part of the organism's interactions with its ambient. In particular, the behavioral distinction made recursively by an observer of his own "states of nervous activity" "constitutes the generation of time as a dimension," and this artifactual dimension is part of his "descriptive domain." It follows that time is part of the observer's "domain of descriptions" and not a feature of the dynamical texture of the ambient or the organism's metabolism.[12]

Concepts for Thinking Texture

Having embedded atomic entities in the more ample notion of texture, let me offer some concepts with which we can think texture from Laozi and Zhuangzi, Gilbert Simondon's account of individuation, nondiscrete topological fields, and David Morris's account of developmental ontology.

留白 (liú bái) Leaving Bare

Let me open with the 道 (Dào)'s approach to partial becoming, indeterminate being. In one of the two principal sources for Daoism, Zhuangzi writes:

[12] "[A] mode of behavioral distinction between otherwise equivalent interactions in a domain that has to do with the states of the organism and not with the ambient features which define the interaction, gives rise to a referential dimension. … [S]equence as a dimension is defined in the domain of interactions of the organism, not in the operation of the nervous system as a closed neuronal, network. Similarly, the behavioral distinction by the observer of sequential states in his recurrent states of nervous activity, as he recursively interacts with them, constitutes the generation of time as a dimension of the descriptive domain. Accordingly, time is a dimension in the domain of descriptions, not a feature of the ambience" (Maturana and Varela 133).

有始也者，
有未始，有始也杯，
有未始，有夫未始，有始也者。

There is the beginning, there is not as yet any beginning of the beginning;
there is not as yet beginning not to be a beginning of the beginning.
(Zhuangzi 15)

Zhuangzi dismisses predicate logic (for instance, arguing about the status of whiteness versus hardness of stone) as "pointless nitpicking debate about similarity" (67) and, I would add, difference. In his essay, "Zones of Indeterminacy: Art, Body and Politics in Daoist Thought," Peng Yu identifies the essential role played by the Daoist notion of 虛淡 (xūdàn, bland, faint, insipid)—from 虛 (xū, empty, weak, virtual, blank, vacant, unfilled, emptiness, void, hollowness, barrenness, diffident)—which "disavows the essentialist territorialization of the attributes as given, unchangeable and passive, and makes their unexpected encounter not only inevitable but also productive" (97).

Daoist philosophical writers, as Steve Coutinho observed in his *Introduction to Daoist Philosophy*, blend pragmatic, poetic, cosmopolitical, and ethical concerns together with what in Western context would be qualified as metaphysical questions. Transposing to creative practice, we have the conceptual and technical tactic of 留白 (liú bái) *leaving bare* areas of a painting to exploit the interplay of there-is and there-is-not. This is quite distinct from the European notions of foreground and background, or of mark and counterspace, in that the regions of bare silk or various intensities of water-diluted or brushed or variably absorbed ink constitute one coincident medium of varying intensity from whose gradients emerge senses of mountains, rivers, trees, figures, houses, and structures in co-constitutive relation with the substrate. The very indeterminacy of 留白 (liú bái) regions engenders and conditions the never-settled change, dynamic, transition, and individuation of more elements of the landscape than could ever be explicitly drawn (Figure 4.3).

Simondon's Individuation and Ontogenesis

Simondon's ontogenetic account of individuation informs much of this essay, but for economy I provide only a restricted account. In *On The Mode of Existence of Technical Objects*, Simondon writes:

Figure 4.3 Guo Xi 郭熙, Zaochuntu 早春图 “Early Spring” (1072). Public domain.

what is determinant and plays an energetic role are not forms but that which carries the forms, which is to say their ground; the ground, while perpetually marginal with respect to attention, is what harbors the dynamisms; it is that through which the system of forms exists; forms do not participate in forms, but in the ground, which is the system of all forms or rather the common reservoir of the forms’ tendencies, well before they exist separately and constitute themselves as an explicit system.

. . .

For the ground is the system of virtualities, of potentials, forces that carve out their path, whereas forms are the system of actuality. Invention is the taking charge of the system of actuality through the system of virtualities, the creation of a unique system on the basis of these two systems. Forms are passive in so far as they represent actuality; they become active when they organize in relation to this ground, thereby bringing prior virtualities into actuality.

. . .

> The milieu plays the role of information; it is the seat of self-regulations, the vehicle of information or of energy that is already governed by information (for instance, the water that is animated by more or less rapid movement, cooling the crank-box more or less rapidly); while the associated milieu is homeostatic, the structures are animated by a *non-recurrent causality*; each goes in its own direction. (Simondon, *On the Mode* 60–1, emphasis added)

This non-recurrent causality underlines the importance of understanding ontogenetic movement as a rhythmicity far more ample than unidimensional periodicity. Simondon replaces the hylomorphic account of transcendent forms imposed on formless matter with an account of forms co-articulating with dynamical matter: "Forms interact not with forms but with their background, … the reservoir of the tendencies of all forms even before they had separate existence or constituted an explicit system" (Simondon, *On The Mode* 63). Matter, far from passive and formless, is ever-structuring, and its ever-structuring constitutes the very operators that structure matter. Simondon accounts for the upwelling of individuation not as the application of an ideal category such as time or geometry to the mute formless stuff of experience, but in terms of what he calls a *disparation* between two orders of the real. Indeed, he provides a decidedly noninformatic, noncybernetic, characterization of information in these terms:

> information is never relative to a single and homogeneous reality but to two orders in a state of *disparation* … is never deposited in a form that is able to be given; it is the tension between two disparate reals, it is the signification that will emerge when an operation of individuation will discover the dimension according to which two disparate reals can become a system; … it is never a given thing; there is no unity and identity of information, for information is not a term; … information is that through which the non-resolved system's incompatibility becomes an organizational dimension in the resolution; information supposes a *phase change* of a system, for it supposes a first pre-individual state that individuates according to the discovered organization; information is the formula of individuation, a formula that cannot exist before this individuation; it could be said that information is always in the present, actual, for it is the direction according to which a system individuates. (Simondon, *Individuation in Light* 11, emphasis added)[13]

[13] One may usefully interpret this "tension" in terms of the immanent stress–energy–momentum tensor inhering in the self-elasticity of dynamical matter–energy–extent.

We will return near the end of this essay to develop a more nuanced interpretation of what could be entailed by the "phase change of a system" in this account of ontogenesis.

Topology

In order to make better sense of Simondon's account it helps to have at hand some notions from nondiscrete topology, notions such as open (set), connectedness, separability, covering, continuous transformation, limit, convergence, and compactness, primordial to metric (quantity) and geometry (form), and primordial in particular to what Brian Rotman called the alphabetic dogma (Rotman) underwriting frame-based theories of consciousness, the central dogma of molecular genetics (Longo and Montévil vii), and category-based or taxonomic approaches to ontology.[14]

The topological concept of open set concretizes the notions of similarity or vicinity. The conceptually deepest aspect of the concretization is that it sets aside quantitative measure and, even more deeply, the very possibility of in-principle-numerical measure. In other words, one does not need to measure anything using a metric (e.g., a distance, whether physical or "abstract") or number to apply this concept of open set. The open set is the most basic notion in point-set topology, but a set is never definable as open in itself; it is always defined relative to a topology, which is a set X of which U is a subset, together with a family of the subsets of X that are declared to be open. Which sets are declared to be "open sets" is up to you, the designer of the topology, provided only that the subsets in this family satisfy the following:

Axioms of Topology

1. If A and B are open, then the intersection of A and B (notated A ∩ B) is open.
   a. The arbitrary union of open sets is open.
   b. The total set X, and the empty set, denoted ø, are both open.

I wish to underline the openness of the concept of open set: given a set X—a universe—there is not necessarily a unique topology. More than one topology

[14] Sha discusses such topological concepts more extensively ("Topology"; *Poiesis*).

may be defined on a given set X. Every set X has at least two topologies. The coarsest topology is the one where the only open sets are X and the empty set ø. The finest topology is the one in which all the subsets are declared to be open.

By definition, a subset C of X is closed if its complement is open in X.

The main lesson here is that the art of a topologist, even at this elementary level, contains a great deal of creative flexibility, that there is no transcendent principle determining a unique topology for every set X.

Mapping

Given topological spaces X and Y we can define *mappings* (also called functions, transformations) from one to another, $f: X \longrightarrow Y$, as an association of elements of X and elements of Y: to every element x in X (written $x \in X$), we associate an element labeled f[x] in Y. The only condition is that the result of applying the mapping f is well defined; that is, that the result is determinate and unique for the given x. A rigorous test: if $f[x_1] \neq f[x_2]$, then $x_1 \neq x_2$ for any x_1, x_2 in X.

A mapping f from topological space X to topological space Y is continuous if and only if the pre-image of any open set in its range space Y is open in its domain space X; mnemonically, "f^{-1}[open] is open"— the pre-image of an open set is open.

Automorphisms

Consider a mapping f from collection, domain, or space X to itself: $f: X \longrightarrow X$. Suppose f is injective: $f[x_1] \neq f[x_2] \leq x_1 \neq x_2$, and that f is also surjective: for all x in X, $f^{-1}[x]$ is not empty. If the domain has some structure, we can consider the mappings f that preserve the structure in some sense. Given that a topology is essentially the weakest and thus most generous structure above the bare category of Set, whose ontological span is arguably too sparse to accommodate lived experience,[15] we can consider continuous mappings on a topological space X. Call these continuous mappings automorphisms. The set of all automorphisms on a topological space forms a mathematical group where the group operation * is composition of mappings: $f * g: X \longrightarrow X$ means apply g and then apply f. We consider this group of automorphisms as the generalization of the concept of

[15] For the full argument, which is nontrivial, see Nirenberg and Nirenberg's "Badiou's Number."

symmetries on X.[16] I will make use of this concept of automorphism to account for sense-making.

Textural Dynamics

As Yu summarized, "留白 (líu bái) is not the representation of the 虛 (xū)-ed ambiguity, it is part of the idea itself. Absent líu bái, things exist in fixed relation to each other and are hence removed from the opportunity for mutual encountering." (This resonates with the Stoics' reconceptualization of predicates as resultants of encounters of generalized bodies.)

This indeterminacy of being—rather than logical status—calls for a consideration of ontological dynamic—a "movement of matter" as envisioned by Lucretius, and Democritus, and elaborated in contemporary terms by Thomas Nail and in a profound way by David Morris with his developmental ontology. This very indeterminacy of texture, this engendering in a way that is not fully determined by what is, entails a dynamicity which is more than merely a spatial physics or a geometric, extensive, movement.

In the following section, I will describe classical "object-oriented" movement, the movement of an atomic entity in abstract relation to another atomic entity, and then contrast that with an alternate account of *textural dynamics*, a variation of gradients of a distributed intensive plenum.

Dynamics: Object-Oriented versus Field-Based or Textural Dynamics

I will give two accounts of dynamics, a hylomorphic one in terms of pregiven entities, and a second in terms of immanent dynamical fields—that is, *textural dynamics*. I will introduce the classical account of the relative motion of gravitating bodies as pregiven objects with mass, and in contrast give an account in terms of a plenum gravity field (Figure 4.4).

Under the classical hylomorphic object-oriented account, suppose we have two bodies with masses m_1 and m_2, at locations p_1, p_2, respectively. (For convenience we name the bodies after their masses.) Newton's law of gravitational force states that the force felt by m_1 due to m_2 is equal (up to a constant factor, which for our purposes I scale to 1) to the product of their masses, divided by the square of the

[16] See Weyl.

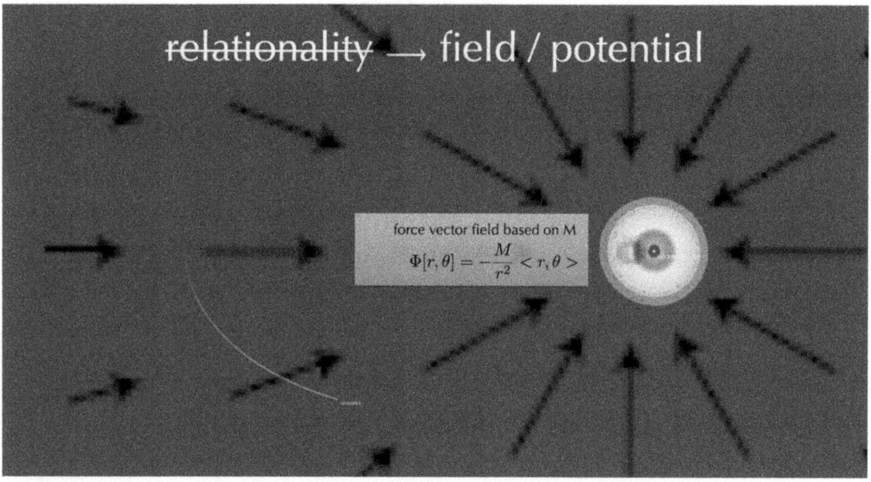

Figure 4.4 "Gravity field due to mass M." Image by the author.

distance between them. The direction of the force acting on m_1 will be pointed toward the center of mass of m_2.

However, this formulation requires having preexisting bodies with determined position and mass.

Let's turn instead to a field-based account. Associated with a body of mass M there is a gravitational field pervading the universe. That gravitational field's strength weakens as the inverse square of the distance to M, and is radially directed to M. At any location a mass m would feel a force directed toward M with strength equal to m * M divided by the square of its distance to M. Notice the subjunctive. The field can be explicitly defined without reference to a determinate "other" mass at a determinate location; a field is a statement of what would happen due to gravitational force if *any* particular mass were to be placed in *any* location. A continuous field holds across a continuum of loci and a continuous distribution of matter. The exact topological concepts of open set and continuity yield nonmetric interpretations avoiding collapse into geometry or physics. It is also important to understand this "field" in a developmental rather than hylomorphic sense. *This field due to M is no less real than M.*

Experience *á la* William James

Note this field expresses a relationality that fully enriches the notion of (dyadic) relation. To see that we begin with William James's empiricist approach to the world in terms of experience as the "primal stuff" primordial to objects and subjects. James helpfully proposes a radical empiricist attitude that takes relations as real as their relata, so that relations are not some abstractions constructed on a universe of *a priori* relation-less atoms. James writes:

> According to my view, experience as a whole is a process in time, whereby innumerable particular terms lapse and are superseded by others that follow upon them by transitions which, whether disjunctive or conjunctive in content, are themselves experiences, and must in general be accounted at least as real as the terms which they relate. (*Essays* 62)

He adds, "subjectivity and objectivity are affairs not of what an experience is aboriginally made of, but of its classification" (James, *Essays* 141).

It's enlightening to read James's essays on experience with *nondiscrete* topology in mind. If we more rigorously interpret what James called "portions" of experience, or experience-processes (and what Whitehead called occasions, entities, droplets of experience), as *topological open sets*, we gain more nuanced traction on experience's dynamics, and relational structuration, including limits, convergence, intensification, disparation. All of which can be expressed topologically with provable theorems on their transductive dynamics and structuration, without reference to metric or quantity, and also without linguistic or graphical glosses that resort to detached metaphor.

A key entry point is to reconsider relation, canonically thought of as a dyadic function of two relata and diagrammed as a figure of thought by unidimensional arc. Under a textural account, entities as well as relations can emerge from and dissolve into contingent dynamics. Furnished with a topological dynamical concept of texture, there is no need to think of relations as dyadic, or even *n*-adic functions. Instead, we can consider fields (which we will see as products of symmetry-breaking differentiations) that co-articulate with continually varying distributions of intensities of experience.

Texture ⊗ Object

I describe the *Meteor Shower* (Figure 4.5), an embodied experiential experiment/experimental experience realized by real-time sound artist Tim Sutton, real-time video artist J. S. Rousseau, and others at the Topological Media Lab. In this installation, a wall mirrors the empty portion of the large space. As A and B

Figure 4.5 "Meteor Shower" installation/experiment, Topological Media Lab. Image by the author.

walk onto the floor, a mist appears on the wall whose density is proportional to the speed of what happens to be moving in the corresponding volume of space. Synthesized sound diffuses into space, modulated by the qualities of movement not of A and B's bodies, but indirectly, by qualities of the diffusion and advection of the mist.

The spatial orientation of the sound matches the distribution of moving matter. At first we hear white noise as A and B leave large clouds hanging on the wall, which dissipate by design according to processes latent in the quasi-physics. A and B improvise movement with their limbs and bodies; then they improvisatorially pick up a long cable that happens to be on the floor and start to swirl it round like a jump rope. The doppelgänger of the rope on the wall starts to create a mist, billowing more in the widest middle of the arc. As the mist thickens further, the latent physics causes portions of the mist to coagulate and condense into lumps. Regions of mist that exceed a threshold density begin to attract one another as more defined gravitating masses, so the forces due to the gravitational field begin to contest the forces due to the "wind" of A and B's movement.

Some of the particles that have been born in these ad hoc dynamics are "voiced" according to sound synthesis processes parameterized by those particles' contingent observables loci and velocities. What starts as a hiss of white noise due to the relatively uncorrelated jumble of turbulent motion resolves into definite chords as the particles begin to move more in a coherent pencil of trajectories.

A vital aspect of this occasion is that none of these sounds are recordings; all the sound and the transition from noise to discernible chordal structure emerges in a locally determinate way from ad hoc activity. The emergent masses—the sounding particles—co-articulate with the optical flow, which in turn co-articulates with the concerted movement of the bodies in the space. Moreover, A and B's momentary, and necessarily intersubjective, intentions emerge in the course of ad hoc activity and the contingent conditions; there is no score, no intention to create any chord, no prior directive for action (e.g., to swing a rope). So the visually and sonically indexed objects of intention are co-articulated with the ad hoc activity, which in turn is conditioned by the sounds and visual cues synthesized by the quasi-physics. For A and B and the others present in that *Meteor Shower* installation, this was a compelling instance of the incessant alternation between structure and transductive operation. And most vitally, *objects of experience* such as rhythms and harmonics emerge as *invariants of experienced textural dynamics*. Therefore, the distinction between objects or subjects versus the indeterminately rich texture of experience is ever-changing and contingent.

Symmetry

As we see in their contingent emergence during the embodied occasions of the *Meteor Shower*, and in a performance of Ligeti's *Poeme Symphonique*, rhythm and relation are not transcendent categories, nor do they preexist the occasion or experience that they condition. Recall the crudest presentation of rhythm as a series of events e_1, e_2, … that are considered equivalent. Instead of figuring this as a unidimensional trajectory, let's repeat the diagrammatic adventure of shifting attention away from a trajectory in the topological space of experience to its indeterminately thick textural complement. We will then consider the automorphisms acting on that topological space of experience and return to consider the invariant subsets that emerge under subgroups of the full automorphism group.

Recall my initial characterization of rhythm as the sense engendered by relative motion of a body across spatial–temporally felt inhomogeneity. Let's develop a more precise way to think about how sense can emerge from the inhomogeneity of the texture of experience. We first illustrate the phenomenon with the plane R^2, calling G the group of affine self-maps on R^2: dilation, rotations (around a point), and translations (see Figure 4.6). A regular hexagonal star shape ⬡, for example, is invariant under the six rotations by multiples of $2\pi/6$ around its center. This means that the rotations by $\pi/3$, $2\pi/3$, π, $4\pi/3$, $5\pi/3$, and 2π, and no other symmetry, will carry every point in ⬡ back into ⬡. Therefore, the subgroup of G that fixes ⬡ has six elements in it. What are the affine transformations that fix a circle? Rotation by any angle around its center will carry the circle into itself, so the subgroup of G that fixes the circle has infinitely many elements in it. We say that the circle is more symmetrical than the ⬡ in the sense that its subgroup of symmetries is larger than the subgroup that preserve the ⬡. Now, what is the subset of R^2, or "shape," that has the most symmetries of all? It's the entire space R^2! No other subset of R^2 will be invariant under *all* the affine self-maps on R^2. I borrow the term *symmetry-breaking* from the physicists, meaning shifting from the invariant of the total group of all symmetries, which is the entire undifferentiated space,

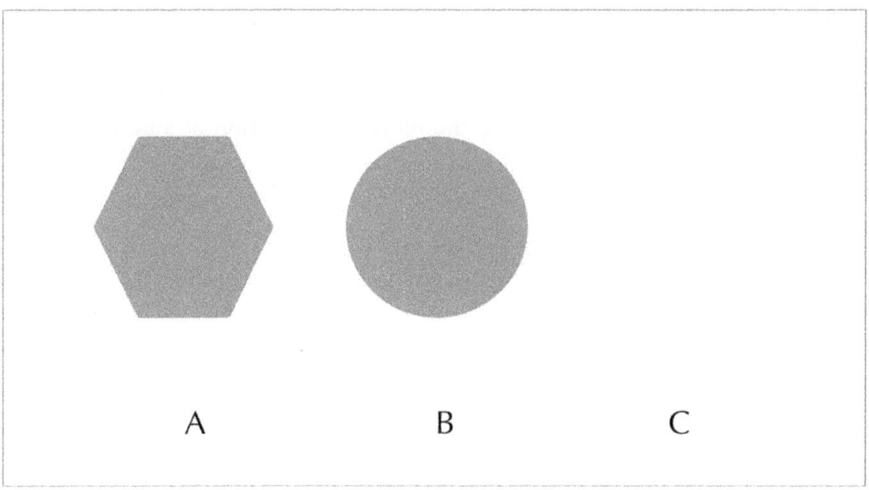

Figure 4.6 "Invariant sets under different subgroups of affine symmetries of R^2." Image by the author.

to the invariant subset associated with a subgroup. A subset distinguishable from the undifferentiated whole is associated with its distinguishing group of automorphisms that leave it invariant as a set.[17]

This simple example featuring affine transformations and regular symmetries is merely a concrete particular to kick-start the proposition, just as Simondon's potent example of crystallization (*Individuation in Light* 75) is a concrete instance but does not model every instance of individuation, especially because every individuation as a process must be singular. In particular, temporal rhythm does not reduce to regular periodicity.[18] And generally, that which is individuated need not have regular symmetry. In Hermann Weyl's words:

> Given a spatial configuration F, those automorphisms of space which leave F unchanged form a group Γ, and *this group describes exactly the symmetry possessed by F.* Space itself has the full symmetry corresponding to the group of all automorphisms, of all similarities. The symmetry of any figure in space is described by a subgroup of that group. (Weyl 45, emphasis original)

However, after illustrating examples of geometric and algebraic symmetries, Weyl elaborates:

> We found that objectivity means invariance with respect to the group of automorphisms. Reality may not always give a clear answer to the question what the actual group of automorphisms is, and for the purpose of some investigations it may be quite useful to replace it by a *wider group*. (Weyl 133, emphasis added)

This resumes our earlier proposition that *objects of experience are invariants of experienced textural dynamics*. To be more adequate to unruly experience, I extend the concept of symmetry to the more ample notion of automorphisms

[17] To be clear, even if $f[\Omega] = \Omega$, $f[x]$ may be different from x. While stirring the soup in its pot, most points are repositioned but the entire mass of liquid remains in the pot.

[18] Simondon criticizes the presumption of principle of individuation that precedes individuals: "It is necessary to reverse the search for the principle of individuation by considering the operation of individuation as primordial" (*Individuation in Light* 3), and elaborates this in his example of the forming of a clay brick in its mold: "The principle of individuation is the singular manner in which the internal resonance of this matter about to take this form is established" (32, emphases original). In context, there's a hacceity to the "principle" of the individuation of this brick, which becomes intelligible under a thoroughly processualist perspective. For a deep inquiry into Simondon's individuation and hacceity, see Malaspina.

on a topological space that I introduced above, and consider the subgroups of automorphisms that leave invariant a subset of that space.[19]

Symmetry-Breaking ⟶ Sense-Making

I take as my guide Noah Moss Brender's core insight that it is the *breaking of symmetry that may be construed as sense*, an insight that he developed carefully in a series of works summarized in "Symmetry-Breaking Dynamics in Development." As Brender put it:

> [P]erception requires an asymmetrical world. Indeed, the more asymmetrical the environment is, the more rich and nuanced will be the sensory feedback it offers to a moving body. This asymmetry is the very texture of reality, which allows us to get a perceptual grip on our surroundings, to "gear into" them with our movements. (Brender 592)

My friendly extension is to substitute a suitably understood concept of *sense* for *perception* in Brender's argument.

We can get some traction on the question "whence asymmetry?" by making more explicit a relation between asymmetry and novelty. One of the most beautiful and profound results of the twentieth century is Emmy Noether's theorem which paraphrases: *to every continuous symmetry there corresponds an invariant (more precisely a conserved current with respect to a flow), and vice versa, to every invariant there corresponds a continuous symmetry*. In light of this bijective correspondence, we can then interpret novelty—that is, a nonconserved current—as a symptom of a deficit of symmetry! This in fact is a deep aspect of the open-ended development of living organisms.

Problematizing the grounding of metaphysics on objects (and subjects), beyond conceiving objects as invariants of symmetries, we consider symmetry-*breaking* as the *emergence* of sense. In Morris's account, sense is

[19] Omitting technicalities, the basic idea is to relativize symmetry to the subgroups H of automorphisms on a topological space M and ask what subsets of M are invariant under subgroup H. Let's call that subset Ω_H the invariant set for H, $h[\Omega_H] \subset \Omega_H$ for all automorphisms h in H. Reciprocally, for a given subset Ω in M, consider the set of automorphisms on M that leave Ω invariant: $H_\Omega = \{h \in \text{Automorphisms}[M] \mid h[\Omega] \subset \Omega\}$. This set is closed under the operation of composition of mappings, thus a subgroup. Guay and Hepburn build toward a mereological theory using groupoids and equivalence classes.

always interpreted processually, and so, as he puts it, developmental ontology addresses not the sense of being but the being of sense, emphasizing "being" in its progressive tense. In his concluding chapter on *developmental ontology*, Morris writes:

> Development does not here mean the development of a determinate program. The determinate character that arises is "to follow," it is engendered in a way that is not yet specified by what is. On the other hand, it does not develop out of a being that is void of determinacy. What there is allows things to develop without yet specifying in advance what will follow. (Morris 230)[20]

Under this view, (the emergence of) sense would not be so much a matter of identifying an object by the subgroup of automorphisms that leave invariant the object in question, but rather the relation of a subgroup to the total group of automorphisms, the "breaking" of the full "symmetry" group, to borrow the physicists' dramatic but imprecise formulation (Figure 4.7).

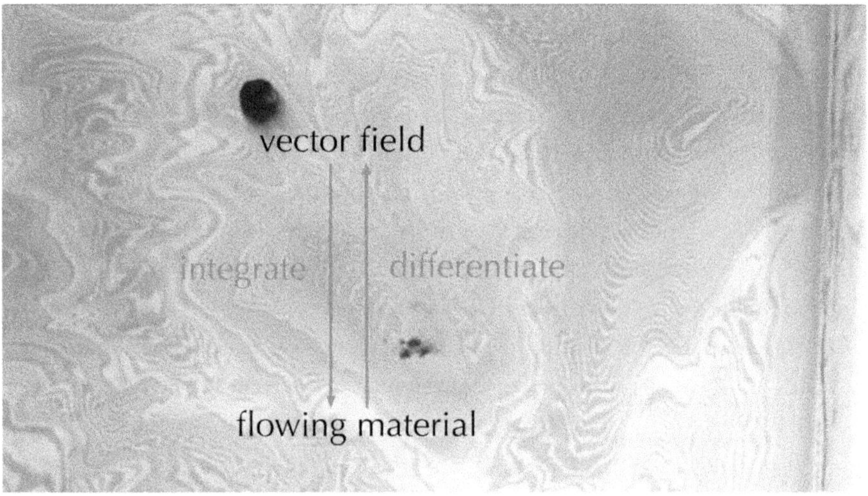

Figure 4.7 "Blowing across the surface ink to create marbling." Image by the author.

[20] Morris takes care to distinguish the transcendent—beyond appearances—from the transcendental: "the *transcendental* is simply an *irreducible* and invariant *structure* revealed or indicated as unsurpassably *manifest within that which appears* or can appear" (17, emphasis added).

Developmental Ontology and Sense

In Morris's account, sense is "manifested within [dynamical] being itself, versus meaning as an ideal or nominalist imposed upon being" (6). He writes, in a lovely figure of thought: "living beings manifest sense and serve as telltales of being, flowing indicators of how sense moves in being" much as the strips of cloth mariners attach to a sail indicate the invisible flows of air over the sail (12). One of the most important aspects of his method is to recognize that one's philosophical reflection is itself a "phenomenon of sense" and, therefore, one's "procedure has to grasp itself as arising within and comporting with sense as a phenomenon *broader than something merely in [oneself]*" (12, emphasis added). Assembling his considerations, Morris arrives at ontological development as a "non-identity of being with itself, such that *being is never fully or already determinately given* ... but can turn out otherwise in ways that are not already foretold [not even!] in terms of, for example, a time or space that could already be foretold" (200, emphasis added). It is in this context that Morris construes time and space as real, rather than as transcendent categories.

Morris's account of the nonidentity of being with itself resonates with Simondon's concept of the transindividual. A superficially simpler way to recognize this is to adopt a thoroughly processualist view, under which all being, all matter, whether physical, energetic, vital (Simondon's term for the living), or social, is dynamical, transforming. In short, matter *is* process.[21] We will see that it is the *emergence* of invariants under subgroups of symmetries that constitutes sense.

Sense-Making in *Steerable Weather*

This more nuanced notion of heterogeneity permits the identification of a nonvanishing differential, and thus a *texture* of orientation(s) that conditions but does not determine the flux of being. And the indeterminate flux of being reciprocally engenders inhomogeneities that constitute nonvanishing differentials. Sense is constituted by these inhomogeneities, or breaks in automorphisms, that *matter*. Suitably understanding forms not as automorphisms

[21] Thomas Nail has persuasively traced the history of this notion of "matter in motion" in a series of works, including: *Lucretius I: An Ontology of Motion*, and *Matter and Motion*.

per se but as particular restrictions to subgroups of automorphisms, we can understand all these operations as taking place *in media res*. Recall Simondon: "Forms interact not with forms but with their background, … the reservoir of the tendencies of all forms even before they had separate existence or constituted an explicit system" (*On the Mode* 63).

Experience 5: Steerable Weather

Equipped with these considerations we turn to the *Steerable Weather* (Figure 4.8) experiential environment, and consider the nonisomorphic but reciprocally conditioned textural dynamics concurrently at play: (1) visual densities, (2) simulated atmospheric physics, (3) a developing field of spatialized sound, and (4) partially intentional corporeal movement.

As you and fellow visitors walk across the floor, its initially featureless expanse condenses into whorls of visual density proportioned to your presence and motion. The latent physics of the simulation computes the advection of quasi-physical water, heat, and air, yielding continually varying distributions of humidity, temperature, and pressure that appear as visual gradients and sonic fields orchestrated in endless variety by real-time video and sound artist-programmers. By turns the sonic fields can sound like timbral washes, polyphonic harmonics, or clusters of bell-like tones arpeggiated not as direct

Figure 4.8 "Weather simulation steered by whole-body movement," Synthesis Lab. Image by the author.

functions of arm and leg movements but interpolated by the textural quasi-physics conditioned but not determined by intentional movement. But there is no causal pipeline from A to B to C. Instead, the inhabitants' relational activity co-articulates concurrently with differentials of the experiential sense of the occasion as its texture varies, together with the actual physical configuration of bodies and architecture, and the computational state of the media synthesis processes.

Deleuze and Sense-Making

We ask, how can sense emerge out of the breaking of symmetries in these textural dynamics? To prepare this, we first look to Deleuze's concept of sense, which steps away from anthropocentric concepts of meaning and object-oriented notions of signification. For Deleuze, sense is neither linguistic nor causal but, as Daniela Voss described it, "an incorporeal added to the object or state of affairs that makes it appear differently, that is, in another aspect or mode" (17). Borrowing from the Stoics, Deleuze interpreted a proposition like "the iron is hot" not as an object with its predicate, but as an *act* constituted by the penetration or mixture of two bodies, such as fire and iron. Sense is, in Voss's terms, an "incorporeal sayable," yet sense is also outside the realm of language.

> Deleuze … draws a qualitative distinction between the realm of causal interaction between bodies or forces, and the realm of impassive incorporeal events or sense. He also distinguishes sense from the realm of language. Being an incorporeal double of the proposition, sense is independent of both affirmation and negation and is also unaffected by the modes of the proposition. (Voss 18)

Now, sense is not, for Deleuze, something deducible from a transcendent logic, but an "effect produced by a specific machinery, a physical, optic, sonorous effect, etc." (Deleuze, *Nietzsche* 137, quoted in Voss 11). In this spirit I introduce the installations as experimental experiences. At the same time, the dynamics at play concurrently span intentional corporeal movement, and poetic equivalences across the simulation: optical flow due to corporeal movement as distributions of hot moist air, color as temperature, height on the scrim as atmospheric altitude, and spatialized sound as turbulence in the generated clouds. So as Voss put it, sense is an "effect that obeys laws of structure that are no longer bound to the depths of existent [actual material causal] forces."

So, taking a developmental interpretation of sense as the incorporeal double that can be said of a situation to change its significance, and adapting Simondon's notion of the phasing of being, we will shift from the causal and epistemic concept of states of mechanical physical systems to *phases* of sense in what Deleuze calls the virtual (Figure 4.9).

First, we review the set-up of a state-based account of a (mechanical) physical system in terms of energy. Let M denote an actual material assemblage, situation-occasion. A decision as to what are the observables then yields a space of possible observables, which we will call the configuration space C. Another decision is the declaration of certain ensembles of configurations as *states*. For example, we can say that a certain range of velocities, water content, temperature constitutes a hurricane, whereas another constitutes a summer squall. If we define a function A: $S \longrightarrow$ R that associates to each state a positive energy, satisfying some basic properties such as nonnegativity and monotonicity, then we have defined what may be considered a potential energy on the space S of possible states of M. The principle of least action predicts that the system will evolve in such a way as to locally minimize the energy incurred in its actual development.[22]

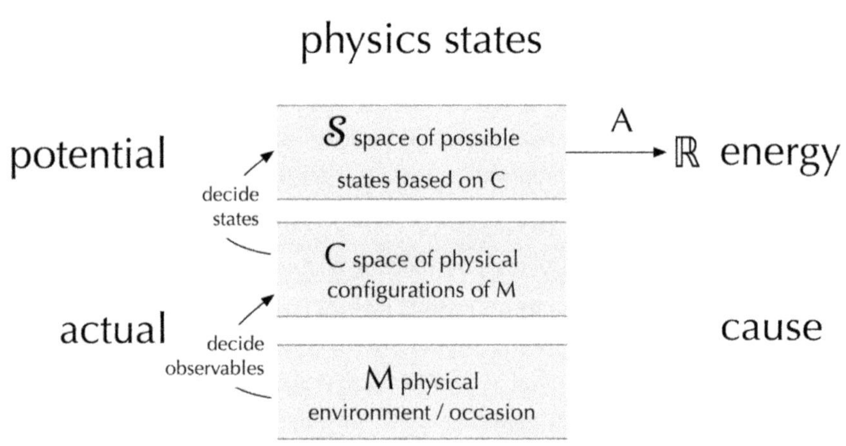

Figure 4.9 "State space of mechanical system." Image by the author.

22 The principle of least action underlies vast swaths of theories, from mechanics and general relativity to Karl Friston's neuroscience, and from Leibniz to Whitehead. For a mathematical history, see Rojo and Bloch.

Interpreting the dynamics of sense calls for thinking not in terms of the *potential* but instead of the *virtual*. And with open-ended development of organisms, as well as social, historical, symbolic dynamics, we can no longer appeal to a universal principle such as the principle of least action (Figure 4.10). (This is a point that for lack of space I take up elsewhere.) Deleuze writes in *Difference and Repetition*:

> The reality of the virtual consists of the *differential elements* and *relations* along with the singular points which correspond to them. The reality of the *virtual is structure*. (209, emphasis added)

And

> [T]he possible ... refers to the form of identity in the concept, whereas the [virtual] designates a pure multiplicity in the Idea. (212)

For Deleuze the virtual is intimately entwined with the Idea as "*differenTiation*" (with a T): the "*determination of the virtual content of an Idea*" (207, emphasis added). Here we need to understand that by Idea, Deleuze characteristically intends the entire material symbolic lived multiplicity. And furthermore, this differenTiation (with a T) is entirely distinct from how differenCiation (with a C) of species and parts is carried out within the actual, within the already-

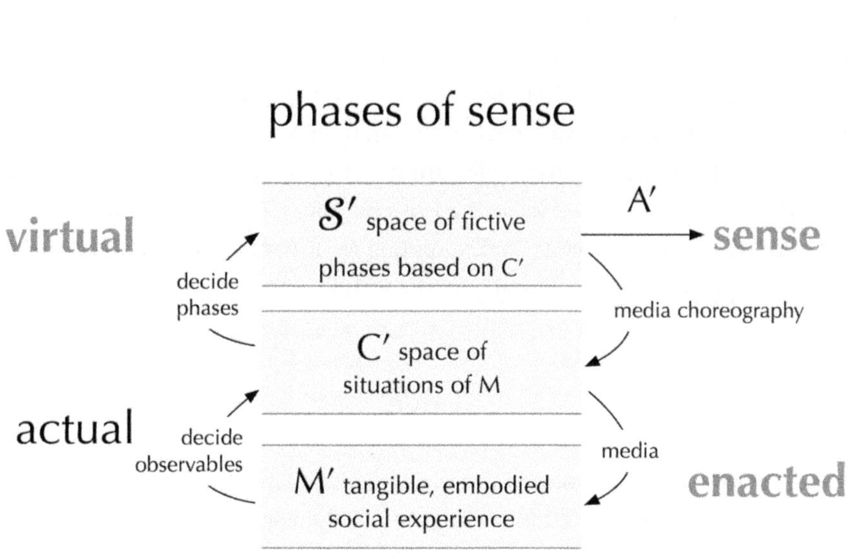

Figure 4.10 "Virtuality: phases of sense." Image by the author.

differentiated conditions of what Deleuze called a problem. As Deleuze put it, "whereas differenTiation determines the virtual content of the Idea as problem, differenCiation expresses the actualization of this virtual and the constitution of solutions" (Deleuze 209). Deleuze gives the example of light, white light. Newton's experiment of running white light through a glass prism and seeing a rainbow of colors emerge on the other side instanced the emergence—the differenTiation, a "determination of the virtual content of an Idea" of color, and of white being composed of all colors (Deleuze 206). The fractioning and distinguishing of reds from yellows from blues amount to the differenCiation of species within the actual. The virtual is opposed to the actual in this sense, but the virtual is always already in the order of the real. This is quite distinct from the possible, which concerns existence.

I close with a proposition more circumspect than understanding, with what Wittgenstein might describe as the dawning of an aspect, a sensation like "that of a light quick intake of breath" (§151). Substituting the concept of the virtual in place of the potential I détourne the physicists' state-based diagram of potential energy animated by the principle of least action. Adapting Deleuze and Guattari's vocabulary (484–93), we have in place of the physicists' physico-energetic manifold the experiential multiplicity of the textures of the symbolic, the social, the affective, which for convenience I will call M′. A decision as to what is observable or discernible yields, or better, co-articulates the situations of M′. This decision process is immanent to the dynamics of M′. Another decision yields the determination of what we can regard as phases in the regime of sense. (Less precisely but more colloquially we can call these phases "fictive.") The processes of differenTiation can then be understood as restrictions ("symmetry-breaking") to particular groups of generalized automorphisms on the virtual phases of sense. Sense then can be understood as the emergence of distinctions among invariants corresponding to disparate restrictions of groups of automorphisms acting texturally on the virtual.

Coda

This essay mirrors the volume as a whole in conceptualizing rhythm as a concept that traverses disciplinary domains sparking fresh insights in its wake. What this essay contributes is a thickening of rhythm as a textural phenomenon yielding a textural account of sense-making. It is indebted to the many adventurous performers, artists, theorists, writers, engineers, and activists affiliated with the

Synthesis Center and the Topological Media Lab who have over the years created media techniques and installation-occasions that co-articulated the conceptual inquiry into sense-making. I hope scholars and practitioners in rhythm studies will carry out further inquiries along adventurous lines inspired by these propositions.

Works Cited

Brender, Noah Moss. "Symmetry-Breaking Dynamics in Development." *Phenomenology and the Cognitive Sciences*, vol. 16, 2017, pp. 585–96.

Bryan, George H. "The Governing Equations for CM1." Boulder, CO. https://github.com/NCAR/CM1/blob/main/docs/cm1_equations.pdf.

Chapman, Owen, and Kim Sawchuk. "Research-Creation: Intervention, Analysis and 'Family Resemblances.'" *Canadian Journal of Communication*, vol. 37, 2012, pp. 5–26.

Coutinho, Steve. *An Introduction to Daoist Philosophies*. Columbia UP, 2014.

Deleuze, Gilles. *Difference and Repetition*. Translated by Paul Patton. Columbia, 1994 (1968).

Deleuze, Gilles, and Felix Guattari. *A Thousand Plateaus: Capitalism and Schizophrenia*. Translated by Brian Massumi. U of Minnesota P, 1987 (1980).

FQRSC. "Research-creation Support Program." 2005. www.gouv.qc.ca/wps/portal/pgs/commun/droitsauteur?lang=en.

Guay, Alexandre, and Brian Hepburn. "Symmetry and Its Formalisms: Mathematical Aspects." *Philosophy of Science*, vol. 76, no. 2, 2009, pp. 160–78.

James, William. *Essays on Radical Empiricism*. Harvard UP, 2003 (1912).

James, William. "A World of Pure Experience." *Essays in Radical Empiricism*. Longman Green and Co., 1912, chapter 2, pp. 39–91.

James, William. "Does 'Consciousness' Exist?" *Essays on Radical Empiricism*. Harvard UP, 1912, 2003 (1905), chapter 1, pp. 1–38.

Ligeti, György. *Poème Symphonique for 100 Metronomes*. Score, translated by Eugene Hartzell. Fluxus Debris!@art/not art, 1962.

Longo, Giuseppe, and Maël Montévil. *Perspectives on Organisms: Biological Time, Symmetries and Singularities*. Springer-Verlag, 2014.

Malaspina, Cecile. "Pure Information: On Infinity and Human Nature in the Technical Object." *Culture, Theory and Critique*, vol. 60, nos. 3–4, 2019, pp. 205–22.

Maturana, Humberto R., and Francisco J. Varela. *Autopoiesis and Cognition: The Realization of the Living*. Boston Studies in the Philosophy of Science. Vol. 42. D. Reidel, 1980.

Mechtley, Brandon, et al. "Enactive Steering of an Experiential Model of the Atmosphere." *Virtual, Augmented and Mixed Reality: Interaction, Navigation,*

Visualization, Embodiment, and Simulation: 10th International Conference, VAMR 2018, held as part of HCI International 2018, Las Vegas, NV, USA, July 15–20, 2018, Proceedings, Part I. Elsevier: 2018, pp. 126–44.

Michon, Pascal. "From Rhuthmós to Rhythm (7th–4th Centuries BC)." Rhuthmos. www.rhuthmos.eu/spip.php?page=spipdf&spipdf=spipdf_article&id_article=1909&nom_fichier=article_1909.

Morris, David. *Merleau-Ponty's Developmental Ontology*. Northwestern University Studies in Phenomenology and Existential Philosophy. Northwestern UP, 2018.

Nail, Thomas. *Lucretius I: An Ontology of Motion*. Edinburgh UP, 2018.

Nail, Thomas. *Matter and Motion*. Edinburgh UP, 2023.

Nirenberg, Ricardo L., and David Nirenberg. "Badiou's Number: A Critique of Mathematics as Ontology." *Critical Inquiry*, vol. 37, summer 2011, pp. 583–613.

Rojo, Alberto G., and Anthony Bloch. *The Principle of Least Action: History and Physics*. Cambridge UP, 2018.

Rotman, Brian. *Becoming Beside Ourselves: The Alphabet, Ghosts, and Distributed Human Being*. Duke UP, 2008.

Sha Xin Wei. "Poetics of Performative Space." *Cognition, Communication and Interaction: Transdisciplinary Perspectives on Interactive Technology*. Edited by Satinder P. Gill. Human–Computer Interaction Series. Springer, 2008, pp. 549–66.

Sha Xin Wei. "What Is Research in Contemporary Art." Topological Media Lab, Concordia University; Synthesis Center, Arizona State University, 2008.

Sha Xin Wei. "The Atelier-Lab as a Transversal Machine?" *Revue française d'études américaines*, vol. 128, special issue on Les études américaines à l'ère des médias numériques, no. 2, 2011, pp. 62–78.

Sha Xin Wei. "Topology and Morphogenesis." *Theory, Culture and Society*, vol. 29, nos. 4–5, 2012, pp. 220–46.

Sha Xin Wei. *Poiesis and Enchantment in Topological Matter*. MIT P, 2013.

Sha Xin Wei, and Garrett Laroy Johnson. "Rhythm and Textural Temporality." *Rhythm and Critique: Technics, Modalities, Practices*. Edited by Paula Crespi and Sunil Manghani. Edinburgh UP, 2020, pp. 101–23.

Simondon, Gilbert. *On the Mode of Existence of Technical Objects*. Translated by Cecile Malaspina and John Rogove. Univocal Publishing, 2016.

Simondon, Gilbert. *Individuation in Light of Notions of Form and Information*. Translated by Taylor Adkins. Posthumanities. U of Minnesota P, 2020.

Synthesis. *Becoming Cloud Installation*, https://vimeo.com/synthesiscenter/clouds.

Synthesis. *Candle-Lighting, Einstein's Dream*, https://vimeo.com/tml/candle.

Synthesis. *Steerable Weather Installation*, https://vimeo.com/synthesiscenter/steerableweather.

Topological Media Lab. *Meteor Shower Installation*, https://vimeo.com/tml/meteorshower.

Voss, Daniela. "Deleuze's Rethinking of the Notion of Sense." *Deleuze Studies*, vol. 7, no. 1, 2013, pp. 1–25.

Weyl, Hermann. *Symmetry*. Princeton UP, 1980 (1952).
Wittgenstein, Ludwig. *Philosophical Investigations*. 3rd ed. Edited by G. E. M. Anscombe. MacMillan, 2002 (1958).
Yu, Peng. "Zones of Indeterminacy: Art, Body and Politics in Daoist Thought." *Theory, Culture and Society*, vol. 33, no. 1, 2016, pp. 93–114.
Zhuangzi. *Zhuangzi: Basic Writings*. Translated by Burton Watson. Columbia UP, 2003.

Part Two

Social Pulsations

5

Arrythmia: Capitalism's Destruction of Rhythm and Relation

Anna Gibbs

This chapter charts the destruction of rhythm and *remaking* of relation in a series of interrupted, *arrhythmic* gestures, attempting—with the help of three "companion objects"[1]: a film, a novel, and a wonderful book about the "war on sleep"—to uncover a horizon beyond collapse. Rhythm and movement bring worlds to life, a state in which everything exists in relation to everything else, forming complex patterns, such that the chaos of proliferation is only apparent. It takes the very different ways of knowing developed by human beings over millennia or centuries by indigenous cultures and by Western science to allow us to look, to listen, and to feel more deeply into this life. For Indigenous Australians, rhythm is everywhere implicit in life: at once a mode of being, a way of life, a series of practices, a conjuring, through which knowledge of "country" is both called up and recalled (encompassing generations of humans and their living environment, including all that Western conceptions of "landscape" treat as inanimate). Indigenous cultures more broadly tend to create relations of "sympathy" and "syntony" with the world,[2] with rhythm serving as the essential element in this crucial synchronization of human life with the natural world.

For Western philosophical and scientific thought, however, rhythm has been constructed as a more isolated object of knowledge, able to be cut out of the world for study by different disciplines in different ways, even if it is also conceptualized as a ubiquitous form of organization or emergent order, palpable everywhere in the natural world. Suzanne Langer, for example, conceives it as "[t]he most characteristic principle of vital activity. All life [she writes] is

[1] The term modifies Donna Haraway's concept of "companion species" (4).
[2] The terms are Berardi's, and he argues that we now need to create new forms of them (*Breathing* 31).

rhythmic" (126). Rhythm animates the world, or perhaps we could say that the animation of the world is organized by rhythm.

As Maxine Sheets-Johnstone (375) writes,

> Animation [is] the foundational dimension of the living. Everything living is animated. Flowers turn toward the sun; pill bugs curl into spheres; lambs rise on untried legs, finding their way into patterned coordinations. The phenomenon of movement testifies to animation as the foundational dimension of the living.

The phrase "patterned coordinations" asserts that rhythm organizes movement, and the rhythms of animation are encompassing. Rhythm not only makes things palpable, apprehendable, comprehensible to humans and doubtless in different ways to other animals and to plants but also creates the complex relations that bring the natural world into being. Rhythm entrains living things, brings them into relation with other things, making for complex processes in which our habitual distinctions between organisms (or between organisms and environments) may be confounded. It creates worlds in which different forms of life are emergent rather than given in advance as things or objects able to be simply or inconsequentially carved out of their environments or *umwelten* for human usage. Rhythm produces emergent form and confers, if not always absolutely strict order per se (Dewey 156), an organization of movement in time. That is to say, it makes *patterns of change* in both time and space.

We too live within rhythms and are animated by them. Rhythm is what binds things and actions, time and space: it is what holds all these together for humans. Think, for example, of the unique rhythms of places, those unique conjunctions of natural and human-made environments, worlds and weathers in which we adapt and in which we intervene by the very process of adaption, accommodation, and remaking to suit ourselves. We create our own rhythms to animate unruly life to our own ends: for example, architectural rhythm orchestrates human movement in relation to space, patterning everyday life in ways we feel without necessarily being consciously aware of it, but also making more palpable our relations to power and authority as these are embedded in the architectures of our institutions, whether instantiated in castles or courthouses. In this way "rhythm enters into the lived," as Henri Lefebvre writes (77), so that we apprehend it corporeally before we know it intellectually.

This is why aesthetic practices so often have rhythm at their heart. As aesthetic theorist Roger Scruton writes, in the experience of listening to music, to hear rhythm "is to hear a kind of animation" between beats that "bring each other into being, respond to one another, and breathe with a common life" (35–6). Rhythm has a strongly affective, even compulsive, component frequently impelling us to motion, whether in dance, or religious practices such as whirling, that, at its most intense, can seem to take us outside of ourselves, beyond the ordinary dimensions of individuality and into a transcendent—depersonalized—realm.

Rhythm (as repetition) is an enabler, making it possible to keep going, as when walking we fall into rhythms that enable our bodies to get us where we are going while our mind freely wanders its own more circuitous routes, or in more organized forms, as when marching to the beat of a drum or hammering a nail. Gertrude Stein knew that rhythm in writing was a means of going on, taking you with it, and rhythm both generates the expectation of continuing even when instantiating it. Rhythm in Stein's writing is processual rather than representational: it emerges from the writing practice itself, whose own routines and regularities shift over time as one phase or mode of her work, eventually creates openings into others. Rhythm creates a certain "being in sync," both within us and our activities and between ourselves and others as it brings choirs or *corps de ballets* or armies or workforces into unison. Rhythm has the power to pull us into entrainment with others, giving rise to relation as it draws us into sympathetic or mimetic rhythms with others.

In The Mood for Love

There can be few more dramatic stagings of the relational work of rhythm than Wong Kar-Wai's film *In the Mood for Love*, which explores the way rhythm organizes worlds (including the aesthetic practice of cinema itself) through the flow of comings and goings within a social milieu constrained by its environment of interior spaces instantiating custom, convention, and propriety in a very particular Hong Kong subculture of the 1960s. Here the two main characters come gradually into sync with each other, and progressively more isolated from the milieu in which they had previously been indistinguishable, much as animals—seahorses, insects, tigers—may be so completely at one with their environments that they are rendered imperceptible. This aspect of relational milieu is emphasized visually throughout the film in the way the colors and patterns of the fabric of the female lead's clothes mesh with the wallpaper and soft furnishings of the apartments in which people gather to talk or play mah jong.

"*Prends garde*," says the anonymous epigraph to Roger Caillois's paper, "Mimicry and Legendary Psychasthenia," "*à jouer au fantôme, on le deviant.*" Or, in English, "Be careful: when you play at being a ghost, you become one," an epigraph that would have served equally well for the film. Mrs. Chan (Maggie Cheung) and Mr. Chow (Tony Leung) discover that their spouses are having an affair, and in trying to understand how this has come about, they mimic the interactions of the couple as they imagine them in the beginning of the affair.

When we first see them, the restless dissatisfaction of Mrs. Chan and the yearning sadness of Mr. Chow complement each other as they come and go, passing each other almost wordlessly in the badly lit spaces of passage—stairs, corridors, backstreets—that form their habitat. Their comings and goings are subject to a shared rhythm that isolates each of them from the rhythms of their respective environments, creating a complicity between them. This is signalled musically by the waltz theme which comes to belong to them: "Yumeji's Theme."[3] From the moment their spouses' affair is revealed, the protagonists' formal similarity is explicit and explicitly recognized by them. The musical theme changes to the first of a series of rumbas.[4] After dinner, walking together in the street absolutely in step and in synchrony, she says to him, "I wonder how it began" (between their spouses)? He responds with mimicry: "Shall we stay out tonight?" To which she replies, "My husband would never say that." They are out of sync with each other at that moment.

Gradually, their rhythms become again more attuned, and when next we see them walking together the game is more extended, becoming increasingly so as their encounters grow more numerous. It becomes less and less clear when one of them begins to talk who the interlocutor is (absent spouse or present interlocutor) and whether the words spoken are "real" or reenacted. This ambiguity, used by the characters as the distinction, is undermined for the audience. Mr. Chan, whom we hear but only see from behind, and similarly Mrs. Chow disappear from the scene, as their selves are entirely replaced by the performances of their spouses.

The synchrony between Mrs. Chan and Mr. Chow becomes more marked. Eventually, in the privacy of his studio, where she now goes frequently, the camera circles around them, creating an intimate and fluid space in which

3 This theme has a quick, restless tempo counterpointed by a plaintive emotional melody.

4 Rumba has a slow, romantic tempo to which one dances on the second beat, the hesitation over the pronounced first beat signifying (at least in this context) the hesitation characteristic of the early and uncertain stages of love and courtship.

we hear almost nothing of what is spoken. Instead, we read their bodies: they smile more, look at each other, engage more openly and more spontaneously than previously. Here they are in sync: her speech and her laugh are formally matched by the rhythm of his gestures. When we next hear their dialogue, they are rehearsing her demand to know from her husband's own mouth that he has a mistress: but at the same time the question suggests itself to her *à propos* of Mr. Chow. At what point does imitation or acting the part of an unfaithful spouse mean that the actors have become unfaithful and untrustworthy themselves? When she's overcome with real emotion, he says to her "This isn't real. It's only a rehearsal." But by this time the rehearsal has become reality, and the love between them, although they never explicitly act on it, has become real.

The "mood" for love—the desire for and disposition toward it—has cast its spell only over the protagonists and is mirrored by the cinematic enchantment of the audience. The creation of this mood depends on more than an identification (in the psychoanalytic sense) with the characters: the location of the story in the past (the 1960s), the exoticism (perhaps especially for Western audiences) of the setting: the décor, the dresses, the music, the cigarettes, the music which moves us, all exert a nostalgic pull, a mood for love, to be thwarted in us as in the characters. This pull and its thwarting that are the point of the film: our mood for love is interrupted by a reminder of historical exigency. We have been depersonalized by the film, drawn into a space/mood: we are "no longer the origin of the coordinates, but one point among others; … dispossessed of [our] privilege and literally *no longer [knowing] where to place [ourselves]*" (Caillois 70, original emphasis).[5]

In the depersonalized multimodal space of cinematic mimesis our synesthetic capacities are heightened, for human mimicry depends for its efficacy on *inexact* repetition and the translation enabled by it, that is, on the capacity for cross-modality, where something presented in one sensory modality (for example, the rhythms of movement) is replicated in another (for example, the rhythms of sound). Finally, though, the film ends, and, like Mr. Chow, we plug up the hole opened within us by the filmic experience and return to a vantage point from which we can see, albeit in retrospect and at the expense of the extraction of our individual bodies from the shared and porous space of spectatorship. In its

[5] This is Janet's "legendary psychasthenia": a disturbance in relations between personality and space. Caillois points out that this capacity for confusion characterizes the scientist, citing the proliferation of represented spaces in contemporary science by way of evidence—but it is also the attitude that characterizes cinematic spectatorship.

beautiful evocation of the rhythms of the social world as habit contingent on habitat—here the apartment complex and the neighborhood such as they were in 1962—the lives of the protagonists seem a little like a human echo of that of the camouflaged insects, indistinguishable from their environment, of Caillois's famous paper.

This echo creates a certain distance from the world it both portrays and observes, inscribing it with the affect of something lost. It is not by chance, then, that *In the Mood for Love* is also everywhere saturated with implicit reference to the existence of a wider geopolitical world at a moment of upheaval. Set amid the refugee Shanghainese community in Hong Kong and made three years after the British handover of Hong Kong to China in 1997, the film speaks *of* a lost world replacing yet an earlier lost world, but *to* the transformed and increasingly anxious present of the world of its making. This transformation encompasses the displacement of populations after the cultural revolution and in the wake of the Taiping Rebellion and the Sino-Japanese War, further displacing many of the Shanghainese amid the riots of 1966 (when Mr. Chow returns to the complex in search of Mrs. Chan, whom he does not find) and also pointing to the threat of further disruption to come, as the year 2046 (supposed to mark the end of the "one country, two systems" policy, and also the number of Mr. Chow's apartment, as many commentators have not failed to notice) hovers over the film as a threatening horizon.

Clocks are everywhere in the décor, pointing to time as an inescapable condition of a moving milieu. The camera takes in radios, lingers on the telephone, surveys all the accouterments of Hong Kong modernity on the 1960s. This is not mere nostalgia but a site and sign of the particular technological, social, and political change that has already disrupted the rhythms of the characters' previous lives and now shapes their current way of life within the wider context of the region (taking in China, Hong Kong, and Cambodia) referred to by the film. It functions as well as an implicit suggestion that more change is to come, speaking to the impermanence—even precarity—of all the routines of everyday life in the modernity of colonialism's wake.[6]

From the vantage point of the time of writing—Sydney 2025, after globalization, after the advent of the digital world of financialization and social media, after the outbreak of new wars in Africa and Ukraine, and after genocide in Gaza

[6] See Audrey Yue for a detailed and brilliantly illuminating discussion of all these themes, encompassing modernity, pop culture, and the *mise-en-scène* of the film.

and in the midst of climate crisis—disruption and displacement seemingly have become a permanent feature of life across the globe. The sense of worlds and times to which we can never return as well as the anxiety of anticipation that pervades Wong's film seems to resonate ever further from the very particular situation he addressed, until it reaches us all in the distinctiveness of our own heres and nows, the local conditions whose particular rhythms are threatened by both global political and larger planetary forces.

Once the rhythms of the natural world—circadian, seasonal, animal migrations, flowering or leaf loss of plants—seemed to provide a constant, able to be taken for granted, a background of worlds in sync, whatever disruption occurred to particular social and cultural rhythms. Even if these rhythms could be disrupted by "freak" weather events, things would stabilize, and the steady background rhythm would resume. Climate catastrophe has shattered this comfortable illusion, as tipping points inaugurating unstoppable feedback loops, with ecosystems collapsing at ever-increasing rates. Extinctions underline the new unpredictability of events as consequences concatenate and change now seems to happen by a process of contagion, as one thing sets off another.

Outbreaks and Contagion

Epidemics of new diseases, resurgences of old ones. Financial contagion. Fires, floods, feelings, hostilities: outbreaks of any kind can happen with or without apparent warning, anywhere and then all of a sudden everywhere. Invisible processes at work below the radar erupt suddenly into immediate, urgent experience. We live in worlds of accelerating contagion. Everything is catchy, and usually not in a good way. Those of us who can, cling to the rhythms, repetitions, and habits of everyday routine, often as a means of shutting out awareness of this very permacrisis. But the fragility of our own routines, and their interdependence on the rhythms and routines of others, are made real for those in the United States and in Europe by the empty shelves where chicken and eggs should be, as avian flu reaches into farmed as well as wild animals, including mammals, a harbinger of wider collapse.

Does collapse have a rhythm? Collapse never happens all at once. A little bit here, a bit more over there, apparently unrelated failures, impalpable to all but those directly affected. Collapse is discontinuous. Collapse is imperceptible until it's almost over. Its interruptions to routine can often be quickly forgotten because the results become normalized and routines change in response to that. Don DeLillo's *White Noise*, published in 1985, precisely delineates this tendency of routine to reassert itself in the face of interruption by crisis as it depicts consumer capitalism a stage beyond the nascence shown in Wong Kar-Wai's film.

White Noise

White Noise is, in many ways, the quintessential novel of late capitalism in a still analogue world from a white American perspective. It takes place in the wake of an unspecified "airborne toxic event" producing an enormous cloud full of "dark and bulky menace." The event is also possibly linked (although "no one had been able to prove it") to an ongoing series of sublime—beautiful and perhaps deadly—sunsets. Giving rise to "broad towering ruddled visionary skyscapes tinged with dread" they are replete with "content, feeling, an exalted narrative life," and seem to be deeply infused with meaning, although people don't know whether to watch in wonder or dread. The sunsets instead produce ambivalence. Likewise, our response to unseasonal beautiful weather now, or perhaps earlier in the pandemic to beautiful images of the coronavirus and its spike protein, produce similar ambivalence. On the one hand the aesthetic pleasure generated by these images seems to affirm life, while on the other they point toward death. The protagonists of DeLillo's novel, Jack and Babette, are preoccupied with death and which of them will die first, although between the series of spectacles provided by the sunsets, the routines of life—getting children to school, supermarket shopping—continue as usual even while Babette secretly tries to dampen her own anxiety about death with an experimental drug called Dylar.

Consumer culture is a constant background of white noise in their lives and offers a never-ending series of new products continually advertised on the ubiquitous media of television and radio, providing at once a source of enchantment while feeding an increasingly obsessive fear about the toxicity of materials and ingredients. It is provocative also of pervasive technological anxiety fed by smoke alarms, news reports of plane crashes, and the authority of data and information (such as medical test results) conveyed by computers. A chemical spill does indeed force the evacuation of the neighborhood to a designated safe campsite, which is soon threatened by a change in wind direction, bringing the toxic cloud overhead and forcing the family's further flight to another city. When they eventually return home, Jack's discovery of Babette's secret addiction, the enforced ending of the clinical trial through which she obtained the drug, and the nature of the deal she did with the director of the trial to maintain access to the drug anyway, all provoke a crisis in which Jack shoots the director and gets shot himself. Once all the fallout is more or less resolved, the novel ends with the shock of entering the supermarket to find that the shelves have been rearranged, and customers are responding by frantically checking use-by dates and scrutinizing the small print of ingredients lists on labels. In the face of this new interruption to routine, Jack resigns himself to a kind of data mysticism in which there is nothing to be done, because it is only the holographic scanners at the checkout that will be able to interpret the occult data they contain. "This is the language of waves and radiation," he says, "or how the dead speak to the living."

White Noise presents the United States at the outset of the digital era, a world in which the analog technologies of *In the Mood for Love* are already beginning to be outflanked. And like that film, DeLillo's novel situates itself in a wider historical context and hints at the contrast with the present of the novel, the change being underlined by the fact that Jack's best friend and colleague, Murray, is a theorist of contemporary popular culture while Jack himself is a leading figure in the historically oriented field of Hitler Studies, a field he founded at the university where he teaches. Hitler Studies seemed hilariously niche when I first read this novel in the 1980s, but now, of course, with the re-election of Donald Trump and the installation of Elon Musk and other cronies prone to making Nazi salutes in the White House, it has a new and ominous resonance for the globally entangled world of the present. Now, however, it's not just that something—a toxic cloud—is in the air. Everything is in the air. It's all up in the air. Or perhaps already all up in smoke. Now, in the age of overt climate catastrophe and the Covid-19 pandemic, we are in life *after* the idea or the experience of the singular event to which disaster might, however tenuously, be traced. We live, rather, in a permanent state of "airborne aftermath" in which all that is solid has been vaporized.

The outdoor atmosphere is now permanently full of various greenhouse and other poisonous gases, and masks are routinely worn on "bad air" days in the worst-affected cities around the world, while indoor air is quite possibly so infused with toxic viral particles we might be contaminated the minute we enter an elevator or an empty room. White phosphorous fills the air in Gaza, and missiles rain down on it while some also streak toward Israel, Iran, and Syria. In Ukraine, too, the air is full of dust and destruction, and the threat of nuclear fallout comes ever closer. Recently bushfire smoke covered much of Greece and Canada, and then Los Angeles, as it did and does again in Tasmania now. This is a world of uncertainty whose fragile familiarity could be shattered at any second. The rhythm of what routine can be re-established is what holds our existence together as it binds days into ongoing life and sutures the wounds of disruption. Meanwhile in places not already openly at war, the "affective atmosphere" (Anderson) is always charged. Ready for a discharge that doesn't come, or comes only in stutters, soon petering out. Rhythm itself seems to be in suspension.

Capitalism, colonialism, and their accelerating consequences for the rhythms of the natural world and our relationship to it are intricately entangled in a complex mass of interconnected circuitries such that cause and effect are confounded. Everything is effect, and everything is affected by everything else. Such circuitries resist biology simply as a metaphor for the social and vice versa: rather, they form flows of heterogenous components—material, affective, tangible, intangible, visible, invisible—in which disparate elements circuits are inextricably interconnected and enmeshed. Anything can connect with anything on any circuit (see Gibbs, "After Effect," "Apparently Unrelated"). I prefer the term "circuitries" to the ubiquitous use of "network," which tends, to my mind, to obscure the uneven degrees of density of connections, as well as the constant mobility and the infinite plasticity of form of this ever-changing mass of circuitries.

The Covid-19 pandemic is exemplary of the disruption of rhythm at every level: as zoonotic disease it is both sign and symptom of the habitat destruction and global heating couple, part and parcel of cascading ecosystem collapse and the runaway climate change which threatens to destroy the seasonal rhythms on which we have always relied. As a communicable disease it places the threat of contagious disease, potentially with extreme consequences for individuals and populations, in the foreground of the everyday social relations whose rhythms it disrupted with lockdowns, even as it provided a new target for biopolitics. As an ongoing global pandemic even beyond the initial "emergency phase" it helps accelerate the digital automation of the world and with it the neurophysiological engineering of affect across various levels of the societies of control, including those most intimate.

Perhaps now, as the next pandemic waits just below the horizon of consciousness, we have tipped from ambivalence into silent dread, and beyond that into denial. If there was once panic about the pandemic, for example, it was mostly pockets of panic buying in the supermarket, not so much widespread panic about the virus itself contrary to some suggestions (e.g., Lee). Nor, unless you are a climate scientist or a committed activist about climate catastrophe, is there widespread panic about that, beyond the immediate responses to specific fire, flood, and hurricane events. Instead, we seem mostly to be paralyzed by our denial, held captive in the sticky web of repetition and routine constituted by the rhythms of everyday business as usual. And now it seems that our frenzied participation in the rhythms of "normality" simply bind us even more tightly in the web of our captivity.

For our part, we often can't even say any more where or from whom we caught the virus. In most places we no longer have access to reliable testing (if we ever had), no reporting, no contact tracing. Contagion itself is no longer an event, singular, but is now a pervasive condition. Said by those in authority (if not by many experts) to be "endemic" (a catchphrase by which is meant simply, prevalent, and by extension, mild), much—even most—of the viral transmission is pre- or asymptomatic. The timing of the onset of symptoms—and the symptoms themselves—alter with each new variant and with the particular constitution of the individual host. Most spread has become cryptic spread. No one can definitively prove they caught the virus at work or in hospital or at school or in the supermarket. We are simply living—or perhaps, in the absence of any meaningful attempt at public health, slowly dying—with Covid-19; as the John Snow Project writes: "Information control has replaced infection control."

At the beginning of the pandemic, lockdowns interrupted the everyday business-as-usual routines, as people who could do so worked from home, attempted to supervise their kids' schooling over Zoom, and perhaps also indulged in learning to bake, or writing a novel, or unaccustomed free-range walking in local streets. While some people responding to a researcher-commissioned Mass Observation survey reported that time dissolved and merged and that lockdown was like being in limbo (Lyon and Coleman), there was nevertheless an opening here for the reassertion of what Lefebvre called "unadministered life," that is, life out of reach of the world of work or governmentality. And indeed, many reported improved sleep (Lyon and Coleman). But even once "unadministered" walking has been increasingly overtaken by Lycra and wellness, by aggressive athleisure wear, and gym subscriptions, and so too cooking by celebrity chefs hawking products to be ordered online.

Reading with all its immersive pleasures has been overtaken by rapid detached scanning of social media or news accessed through it. None of these things offer the *dérive*, still less the *détournement*, practiced by the Situationists. Botanizing on the asphalt—or any kind of reverie and reflection—has been replaced by data. Who has time anymore? In fact the lifting of so-called restrictions (actually public health protections) sold to us as "freedoms," and the end of working-from-home orders and of pandemic payments that raised many people (especially single mothers) out of poverty, actually represents forced returns to offices and wage slavery, and a new step in the destruction of the collective enterprise of postwar (re)construction: the building of institutions such as the Public Service, welfare safety nets, and all manner of physical infrastructure in a world still committed to "the common good" represented by public health.

This latest phase of pandemic response arguably marks a new stage of the "24/7" world of contemporary capitalism that profoundly disrupts the rhythms of everyday life. In an excoriating description of the always on world of the "24/7" such as it was at the time of writing in 2013, Jonathon Crary explains that it "has the semblance of a social world but, [….] is actually a non-social model of machinic performance and a suspension of living that does not disclose the human cost required to sustain its effectiveness" (9). This "always on" world wages war on the very sleep that sustains the rhythms of everyday life. As Crary points out, the "hyper-industrialism" (Stiegler's term) that emerged in the 1990s gave rise to a "mass synchronization" of experience—not at all the same thing as the "shared temporalities in which the co-presence of difference and otherness could be the basis for provisional publics or communities" (51). In contrast with Stiegler, Crary's analysis reaches even further beyond the peacock's eyes or attractors of mass distractions of the entertainment industries drawing us into delivering our data into the "larger and compulsory arrangements within which these elements, and many others, are consumed" (52).

Beyond these kinds of content, "online gambling, internet pornography, and video-gaming" whose "drives and attitudes […] with their illusions of mastery, winning and possession, are crucial models for the intensification of 24/7 consumption" (53). Information and communication systems, neurochemicals, drugs, legal and illegal, all furnish further examples of the means by which the "parcellation and fragmentation of shared zones of experience into fabricated microworlds of affect and symbols" (53) takes place. Meanwhile,

> [c]onventional and older durational units persist (like "nine to five" or "Monday to Friday"), but overlaid onto them are all the practices of individual time management made possible by 24/7 networks and markets (57).

Crary wrote his brilliant little book at the outset of a world now dominated by social media, before social media became the current all-pervasive phenomenon and before the advent of artificial intelligence (AI). Nonetheless, his work is absolutely clear-sighted in its detailing of the destruction of ritual, cyclical repetition, and rhythm by the relentless forces of contemporary capitalism in which less and less time is free from administration and more and more of so-called free time is further commodified and less and less able to provide a source of resistance to the onset of each new development of capitalism.

Moreover, the everyday, according to Crary, "is at once too fugitive, too uncircumscribed to be imagined as a field of counter-practices to the codes and institutions of modernization" (69), despite the fact that "forms of opposition and resistance" have sometimes emerged from it in the past. The pandemic is routinely presented as demarcated and sealed off in time past, yet we are now "post-pandemic."[7] Any emergent shoots of new forms of local community and connection arising out of its early emergency stage have been left to wither away as the global routines of business reassert themselves. "[I]t is in [nature of] the everyday," Crary writes, "to adapt and reshape itself, often submissively, to what erupts or intrudes in it" (69–70). If it is characteristic of the everyday tending to simply absorb whatever has changed into the new normal without any explicit recognition of doing so, it is also clear that the process of absorption is increasingly engineered, as the pandemic response reveals. In the aftermath of the Omicron variant, a series of catchphrases (*mots d'ordre*), striking in their similarity across national boundaries (and all originating in the Great Barrington Declaration or McKinsey advice to governments), began to be crooned. "Personal responsibility" and "freedom" began to become familiar refrains, along with "Omicron is mild," the virus is "endemic," "living with Covid," "children don't get infected," "children don't pass it on," and so on.[8] Governments and corporations alike wanted return to business as usual.

The result has been that it seems normal, now, for people to be recurrently ill, or for healthcare workers in cancer treatment centres to be unmasked around immunocompromised patients. Meanwhile, travel advertisements for exotic (to

[7] But we are not in truth post-pandemic, and nor do pandemics ever really end so neatly. The history of the Spanish Flu is a good example: think, for example, of the increase in heart attacks in the years after it, or of the legacy of narcolepsy still visible in some Spanish and Basque families even today.

[8] Several authors, including Jan Gresil Kahambing and Blair Williams have analysed the return to apparent normality as an exercise in the manufacture of consent by governments and business.

many Australians) European worlds flood social media. What they don't say is that these worlds no longer exist or barely exist except in a state of collapse. Australian screens are once again filled with the familiar enticements of scenic river cruises through various European cities past vineyards and fruit trees and rows of carefree sidewalk café tables and enduring cultural legacies embodied in ancient buildings, while passengers feast on local wine and culinary delicacies. These now appear among media images of towns and villages drowning and all but destroyed under torrents of turbulent water. This is biblical flooding, sometimes in the same places where just a few months ago there was apocalyptic fire (Greece, Portugal, Italy, Slovenia, France, Romania, Austria, Canada, California—and, indeed, Australia), wiping out farmland and poisoning the soils, making it impossible to grow never mind harvest food. And yet life apparently goes on. Even though the ecosystems that once supported all the—doubtless romanticized—routines of everyday life in such places (the cuisines dependent on fresh produce, the vineyards, the bathing in crystalline waters) have all but collapsed and the complex, multilayered rhythms typical of public places in European cities are being increasingly replaced with the orchestrated rhythms of purely utilitarian commuter and consumer passage through the smooth surfaces of corporatized and privatized space.[9]

As in DeLillo's novel, the demands of capital drive technological developments that disrupt and destroy the familiar rhythms of the everyday. Crary identifies the "counter-revolution of the 1980s and the rise of neoliberalism (the cultural consequences of which are delineated in *White Noise*), and with it the marketing of the personal computer and the dismantling of systems of social protection along with the monetization of time as a major point of intensification of the 'assault on everyday life'" (70–1). Crary argues that it is not technology per se, but rather the changing structure of capitalism, that always drives particular technological directions so that the world of financialization, of derivatives and debt, is both made possible by and yet equally demands new developments in digital technology. While endorsing Deleuze's conception of the control society arising out of this (71), he sees (unlike Deleuze) the "continuous forms of control" in which he observed "an additional layer of regulation alongside the still functioning and even amplified forms of discipline" (72). This is borne out today by the end of pandemic "work-from-home" rules and the calls from

[9] In this world, it seems, suggestive environments have arguably become coercive environments in which the SARS-Cov 2 infections that were once to be avoided have now become almost mandatory so as to attempt to engineer the (much vaunted but entirely fantasized) herd or hybrid immunity.

business for compulsory returns to the workplace and the end of flexibility for workers altogether, even though technology makes it entirely possible to work remotely. In tune with Guattari's more recent work, Crary points to an expansion of consumerism in wealthy societies, such that it becomes a nonrefusable, compulsory "24/7 activity of techniques of personalization, of individuation, of machinic interface and of mandatory communication" (72).

Crary views arising from these conditions the ever-more intensified parcelization of collectivity into Sartrean seriality, that is, into "an aggregate of discrete individuals who relate to each other only on the basis of hollow or narcissistic identities" (116). Contemporary social media now both generates and epitomizes this seriality and with it the destruction of the everyday as rhythm as well as routine. Where once the habitual automatisms of the everyday were once structured by the suggestive environments of architecture and urban design as we moved through cities, they are now also increasingly regulated by the algorithmic architectures that more and more drive what now passes for social interaction and social relationship is actually merely transaction. Social media has become a major enabler of surveillance, control, and the extraction of monetizable data that characterize the process of destruction Cory Doctorow calls "enshittification", a term he applied to describe the way platforms suck us in, lock us in, then degrade such utility for us.

Such a process that characterizes the form of contemporary capitalism—or, more properly, what Yanis Varoufakis calls "technofeudalism"—where rent (which is derived from privileged access to things in fixed supply) rather than the profit from investment in production (which, unlike rent, is always open to competition) is the major source of income for corporations and small numbers of very wealthy individuals. Beyond its algorithmic creation of echo chambers through its manufacture and amplification of rage and outrage, its control over who sees what, its use of dark patterns, and its production of addiction to ensure continuous immersion in its relentlessly regulated machinic flow, social media also destroys the rhythms inherent in both the creative forms of montage envisaged by both Benjamin and Eisenstein and those of everyday life.[10] Rhythm depends on the difference (*différance*) generated by interval, but social media

[10] This is not to say that there aren't forms of comfort and community to be found in social media, but it is arguable that, as Crary writes, if "the networks are not in the service of already existing relationships forged out of shared experience and proximity, they will always reproduce and reinforce the separations, the opacity, the dissimulations, and the self-interestedness inherent in their use" (121).

nullifies difference. As Will Luers writes, its continuous feed ensures that "the intervals between images, texts and videos are squeezed, smoothed out and made uniform," such that "[t]here are no authored spaces between Tweets or Facebook posts. No pauses or silences between the graduation picture and the obituary." Instead of entrainment in the rhythms of the world, social media draws us into participation in—that is, becoming a cog in—the circuitry controlled by refrains; that is, by the order words (*mots d'ordre*) or slogans that function as linguistic operators in the abstract machinery of code.

Almost nowhere is the destruction of rhythm so complete as in the university. Once teaching was about relationship; it involved the forging of bio-neural connectedness between teacher and students, attuning and a tuning in which a mimetic component (admiration, aspiration, and emulation) played a major part, along with the turn-taking of discussion or Socratic teaching—it was possessed, in other words, of a certain rhythm. Teaching involved responsiveness to other bodies, tuning in and attuning to, often punctuated by the spontaneous eruption of novelty in thought, after which interruption the rhythm of discussion is restored. For students, education was about the transformation of the self through participation in an intellectual milieu often deeply critical of the world beyond the university while highly engaged with it. Seminars and tutorials required the patience to listen and to wait to take one's turn in speaking, a sensitivity to the social rhythms of the classroom. Such patience is entirely abrogated by the contemporary furnishing of everything at will and "on demand."

Education

University education is now no longer even exclusively synchronous (live lectures, tutorials, seminars), but mostly asynchronous, comprising "of flipped" classrooms on Zoom or recorded video lectures. Students no longer complete degrees within a cohort, as the rhythms of progression are disrupted by students' increasing need to work as they study, and by the morcellation of courses into units of "micro-credentialling" to produce what in Australia are now officially called "Job-Ready Graduates." Activities that once formed part of teaching and learning—like the writing and marking of papers—are now being outsourced to automated software. Just as social media presents a "simulation of life" and Google a "simulation of cognition" (Berardi, *The Second Coming* 130), the university now presents a simulation of education, in which all the functions of the university from education and its administration right through to campus security are increasingly outsourced, as Matthew Kirschenbaum and Rita Raley note, to digital platforms and third-party vendors. They conclude that at the limit the university itself becomes:

> a field of deployment, or platform in the sense that the physical site of the campus is arrogated as a nexus for the capitalized ideological vectors that ramify throughout algorithms and "services" of all kinds, of which the service named higher education is now merely one homologous epiphenomenon. (Kirschenbaum and Raley)

In this respect, far from being an isolated ivory tower, the university has become a microcosm of the wider world: everyday life beyond the university has now also become an epiphenomenon of a data-driven world. Data extraction is the new mining, driving the decisions of government and business alike.

As Berardi explains, power is now a "statistical function emerging from the noise of the crowd"—in other words, from the noise of social media (*Breathing* 27). What this means is that the "future is now being written by the algorithmic chain inscribed in techno-linguistic automatisms." Much human behavior has become one more vector of code. With the scale and speed of new forms of AI working at levels beyond human comprehension and the concatenating existential threats from the revenge of the natural world we now face mean that all models ultimately fail us, making it difficult to see how, where, and with what results we could intervene. When everyday life is devoid of rhythm beyond mechanical routine in which the succession of passionate identifications of selves are reduced to the fixity of identity. As even the rhythms of the natural world are out of sync and the disruption of seasonality is destroying the lives of ecosystems, we must ask whether any restoration of rhythm is now possible.

Since inception of the disastrous colonial enterprise, we have been doing too much acting *on* the world rather than *in* it or in tune with its rhythms, in the much less damaging and indeed actively nurturing ways of Indigenous technologies, invented and adapted over an almost immeasurably longer timescale. This implies a way of being that follows an ecological model, small scale, mobile, flexible (responsive to local conditions and local contingencies) and necessarily slow. In keeping with such an ethos, Berardi acknowledges the dependence of life, and the affective communication that sustains humans in common, on bodies.[11]

[11] "[M]eaning does not lie in things, or in the signs of language. It is generated by the endless shift from one interpretation to the next, from the uncertain and ambiguous exchange of gestures. Desire is the energy that enables this continuous activity of interpretation. Meaning is the effect of affective communication among language agents" (Berardi, *Breathing* 71–2).

Breathing has always been a touchstone for rhythm and the attunement with ourselves and others enabled by it, but contemporary capitalism leaves us no time or space in which to draw breath, still less to seek out syntony, the external rhythms of the world that match our internal ones. For Berardi, such restorative syntony requires an essential remaking of the world, and of ourselves in relation to it. If Sartre's famous image of arrhythmic seriality is a queue in which both selves and sociality languish interminably, Chantal Akerman, even as she shows us seriality in all its bleakness in her film *D'Est* (1993) made after the fall of the Berlin Wall, nevertheless also, as Crary suggests, gives us the merest glimpse of the queue as a space in which tentative community might just possibly open to establish a different rhythm, one allowing the emergence of something new. Now, perhaps waiting rooms, especially clinical ones, might be a particular pandemic hell in which "l'enfer c'est les autres," as in Sartre's notorious image of the lift, but the very existence of an airborne zoonotic virus at least reminds us of the porosity of bodies and beings, and of the dual aspect of this, as at once potential threat and an essential condition of life. Godot is not coming, and out of this strange space of denial that collapse has begun, we need to begin to forge new rhythms for a new world.

Works Cited

Anderson, Ben. "Affective Atmospheres." *Emotion, Space and Society*, vol. 2, 2009, pp. 77–81.

Berardi, Franco. *Breathing: Chaos and Poetry*. Semiotext(e), 2018.

Berardi, Franco. *The Second Coming*. Polity Press, 2019.

Caillois, Roger. "Mimicry and Legendary Psychaesthenia," *October*, vol. 31, winter 1984, pp. 17–32. Originally published as "Mimétisme et Psychaesthénie Légendair."

Crary, Jonathan. *24/7: Late Capitalism and the Ends of Sleep*. Verso, 2013.

D'Est, [film] directed by Chantal Akerman. Icarus Films, 1993.

DeLillo, Don. *White Noise*. Penguin Books, 2016.

Dewey, John. *Art as Experience*. Perigee Books, 1980.

Doctorow, Cory. 2024. https://doctorow.medium.com/my-mcluhan-lecture-on-enshittification-ea343342b9bc.

Gibbs, Anna. "After Affect: Sympathy, Synchrony and Mimetic Communication." *The Affect Theory Reader*. Edited by Melissa Gregg and Gregory J. Seigworth. Duke UP, 2010, pp. 186–205.

Gibbs, Anna. "Apparently Unrelated: Affective Resonance, Concatenation and Traumatic Circuitry in the Terrain of the Everyday." *Traumatic Affect*. Edited by

Meera Richardson and Michael Atkinson. Cambridge Scholars Publishing, 2013, pp. 129–47.

Haraway, Donna. *Staying With the Trouble*. Duke UP, 2016.

John Snow Project. "'Endemic' SARS-CoV-2 and the death of public health." November. https://johnsnowproject.org/insights/endemic-sars-cov-2-and-the-death-of-public-health/2023.

Kahambing, Jan Gresil S. "COVID-19, Conspiracies and Manufactured Consent." *Journal of Public Health*, vol. 44, no. 3, 2022, pp. e465–e466. https://doi.org/10.1093/pubmed/fdab300.

Kirschenbaum, Matthew, and Rita Raley. "AI and the University as Service." *PMLA/Publications of the Modern Language Association of America*, vol. 139, no. 3, 2024, pp. 504–15.

Langer, Suzanne K. *Feeling and Form: A Theory of Art*. Charles Scribner's Sons, 1953.

Lee, Raymond L. M. "Affectivity, Subjectivity, and Vulnerability: On the New Forces of Mass Hysteria." *Subjectivity*, vol. 15, 2022, pp. 18–35. https://doi.org/10.1057/s41286-022-00127-6.

Lefebvre, Henri. *Rhythmanalysis: Space, Time and Everyday Life*. Continuum, 2004.

Luers, Will. "Making and Breaking Space: Rethinking Montage in Digital Writing." *Rhizomes: Cultural Studies in Emerging Knowledge*, no. 36, 2020, online.

Lyon, Dawn, and Rebecca Coleman. "Rupture, Repetition, and New Rhythms for Pandemic Times: Mass Observation, Everyday Life, and COVID-19." *History of the Human Sciences*, vol. 36, no.2, 2023, pp. 26–48. https://doi.org/10.1177/09526951221133983.

Scruton, Roger. *The Aesthetics of Music*. Oxford UP, 1997.

Sheets-Johnstone, Maxine. *The Primacy of Movement*. Benjamins, 2011.

Varoufakis, Yanis. *Technofeudalism, What Killed Capitalism*. Penguin Random House, 2023.

Williams, Blair. "The Politics of 'Letting It Rip': Why Australia went from Zero-COVID to COVID-Central." *Research Handbook on Public Management and COVID-19*. Edited by Helen Dickinson et al. Edward Elgar Publishing, 2024, pp. 72–84.

In the Mood for Love [film] directed by Wong Kar-wai. Block 2 Pictures, 2001.

Yue, Audrey. "In The Mood for Love: Intersections of Hong Kong Modernity." *Chinese Film in Focus: 25 New Takes*. Edited by Chris Berry. British Film Institute, 2003, pp. 128–36.

6

Rhythm in and of Social Interaction: A Study in Intersubjectivity and Power

Chiara Bassetti

> *... all social interactions connected with the musical process: they are doubtless meaningful to the actor as well as to the addressee, but this meaning structure is not capable of being expressed in conceptual terms ... For this very reason it can be hoped that a study of the social relationships connected with the musical process may lead to some insights valid for many other forms of social intercourse, perhaps even to illumination of a certain aspect of the structure of social interaction as such.*
>
> (Schutz 159–60)

Introduction

We have rhythms in our body, we are made of rhythms. We produce rhythms in action, we make rhythms. We perceive rhythms via and through our body, we are continuously remade by rhythms. We co-create rhythms in interaction, we make rhythms together. And this all happens at once. This chapter addresses some of the threads of such an entanglement, with particular attention to social interaction and its aesthetics. In doing so, it focuses on intersubjectivity, meaning-making, and situated power management. It addresses questions such as: Which kind of meanings and forms of knowledge is rhythm specialized to convey and manage in interpersonal interaction? How does that get done? Does the seductiveness of rhythm bear only positive social outcomes or is there a "dark side"? Whereas experimental research that explored the role of rhythm on affect, attitudes, cognition, and behavior (e.g., Mogan et al.) mostly considered dyadic interaction, often of task-based nature, this chapter, grounded in ethnomethodology and conversation analysis (EMCA), focuses on rhythm in—and of—naturally occurring, everyday social interaction, including nondyadic (i.e., group), situations ranging from convivial, to conflictual, to work-oriented.

More specifically, by leveraging both interdisciplinary literature and my own empirical data analyses, I shall argue: that rhythm in social interaction, from primary socialization to more sophisticated practices such as the performing arts, has the capacity to seduce us and builds to intersubjectivity; that rhythm carries relational meaning more generally, and can help us travel along the nonconceptual–conceptual continuum, in music and dance as much as in conversation; that, in fact, rhythm is employed to collectively generate—collaboratively, competitively, or by leader–follower dynamics—distinctive *interactional aesthetics*, and that these are socially meaningful (i.e., they display and put into being given social situations and given social relationships, thereby also allowing expressing and managing power relations and power, including charisma, in situated interaction). It is a "certain aspect of the structure of social interaction as such" (Schutz 159–60) that this chapter aims to uncover—an aesthetic aspect, grounded in rhythm.

Rhythm in Interaction

Interacting Bodies in Rhythms: A Seductive Experience

Several bodily phenomena are rhythmic, such as heart beat or breath (endogenous rhythms). This embodied dimension also concerns movement. In its most basic layer (i.e., alternation between muscular tension and relaxation), movement rhythms affect us: "biting on a pen might indicate the need to focus … curling one's hair or rocking in a sucking rhythm … the need to soothe oneself. Likewise, these rhythms are employed to address needs of others (e.g., soothing a baby; consider that all lullabies consist of sucking rhythms) and usually lie at the implicit level of our experience" (Koch 3). This is where rhythms remake us and allow us to shape others' lived experience.

Many studies across several disciplines have addressed the rhythm of moving in interpersonal synchronization and entrainment, leading to mutual coordination—with or without the aid of external rhythms. Shared exposure to rhythm facilitates interpersonal motor coupling, as observed both in experimental settings (e.g., Lang et al.) and naturally occurring group activities such as unison reading (e.g., Cummins), marching together on drums (e.g., McNeill), dancing together on music (e.g., Bassetti, "Knowing"), or rowing together on talk (e.g., King and de Rond). But people can get entrained in one and the same rhythm of dancing within total silence (Bress), as breathing and moving together can accomplish the same (if not stronger) result as listening together to music—consider, for instance, interpersonal sexual practices (Safron). Further, not only joint activity (Palumbo et al.) but also visual access (Oullier et al.) and co-presence allow for interpersonal synchronization, even when we are *not* set to accomplish a task together or to directly communicate (Golland et al.). In short, we are hard-wired for rhythmic coordination (Collins, *Interaction*; *Violence*), and we have developed several tools (e.g., music, drumming, chanting) to further ease the task (e.g., Savage et al.).

It is not only about effective coordination in action; experiencing rhythm together is transformative, influencing our attitudes (Kraus 121). Evidence within the cognitive and neurosciences built a strong case for prosocial dispositions being favored by synchronicity and eurhythmicity (e.g., Mogan et al.; Vicaria and Dickens). "Entrainment facilitates social bonding" (Gill 118) and cohesion. Interpersonal synchrony has been found a driver of affiliation and likeability between co-participants, at the dyadic (e.g., Hove and Risen) and group level (e.g., Gordon et al.). We also seem to "read" movement synchrony, for instance in walking, as marking a social unit (Lakens and Stel). The same, I suspect, lies behind the concept of "*corpo di ballo*"/"*corps de ballet*" (Bassetti, *Genesi* 81–90). Further, an experiment by Cirelli and colleagues showed how the pleasurable experience of the synchrony of motor and auditory stimuli the adult allowed when bouncing babies on- rather than off-beat with music playing led to cooperative behavior afterwards. In previous work (Bassetti and Bottazzi 465–71), analyzing naturally occurring adult–newborn interaction, I showed not only how the creation of rhythm—by moving toys, or by beating glasses one against the other in a progressively faster and complex rhythm—engages and ent(e)r(t)ains the baby, but also how the infant selects, on such a basis, an adult among others to whom devoting attention. "[A]s Plato says (cit.), it is the pleasurable perception of rhythm … that causes us to move" (459). Rhythm leads attention, it has a seductive power.

Overall, rhythmic alignment fosters pleasure, attention, and prosocial behavior. Indeed, dancing together, singing together, cheering together—in short, acting together on and with rhythm, the shared experience of rhythmic action—is powerful, it causes us to move, as Plato maintained, while it "moves" us, as with music (Vuoskoski et al.). Relationships are continuously (re)produced through this process; they are made of this matter, one could say. Sociologist Randall Collins's (*Interaction*) concept of *interaction-ritual chains* is precisely to the point. Moreover, although the idea of the "madding crowd" is a myth (McPhail), mass entrainment phenomena such as those observable at huge concerts, rave parties, religious ceremonies, sport events, and the like engender what Durkheim (405, 441) defined "collective effervescence," which in turn is capable to generate a sense of togetherness, a "We-relationship" (Schutz 173).

Rhythm and Meaning in the Performing Arts: The Nonconceptual

Rhythm not only allows us to share embodied and affective states, but also meanings, and particularly social meanings in a nonconceptual format. As I aim to show, when nonconceptual knowing is concerned, rhythmic material is crucial, alongside—and prior to—objectivation tools such as inscriptions and verbal accounts (cf. Liberman, "Objectivation"). Whereas such tools are methods for explicitly sharing meaning by making it conceptual and for stabilizing knowledge, rhythm is one of the means we have for the nonconceptual interpretation of experienced reality and for implicitly *sharing unobjectivated knowledge*. In this section, I take the case of performing arts to show how rhythm is a potent signifier, and a social signifier. Performing artists can encode meaning into nonconceptually expressed forms such as music or dance, but they also travel the other way round when the aesthetic material of their art needs to be taught. In the former, both action rhythm and interactional rhythm are tools for meaning-making; in the latter, rhythm in the realm of talk or gesture carries meaning for describing and prescribing rhythm and other aesthetic qualities in the realm of "the performed"—a tool for meaning sharing.

I start from the first "journey" and from music. As Schutz observed, the meaning structure of musical processes cannot be expressed in conceptual terms, yet it is meaningful both to its actors and audience. Music is an empirical example of the sharing of pre-predicative meanings. The latter include relational meanings. Improvised music, and jazz in particular, has been read as conversation in several sociological analysis. In another disciplinary domain, Aucouturier and Canonne see music as "perceived as the sonic trace of *social relations* between a group of real or virtual agents" and talk of "full-fledged social scenes heard by the listener in the music itself" (94). They provide experimental evidence of "the capacity to decode social cues from acoustically-expressed *coordination* of non-verbal material" (99, emphasis added), "processed in complementarity to the single-channel, 'prosodic' cues contained in the expression of each individual musicians" (104). Whereas this capacity appears to be of us all, musically trained and not, encoding relational meaning (categories of interaction such as domineering, disdainful, or caring) into improvised duets was a task assigned to professional musicians in the considered study. In the next section, I show such meaning-making holds not only for culturally sophisticated nonverbal material such as music, nor only for mundane nonverbal practices like backchannelling (106), but also for talk in interaction. In their closing remarks, Aucouturier and Canonne consider speech, but their conclusion is contrary to what I shall argue:

> In speech, one is expected to give floor when agreeing and take the floor when contradicting ... Here, on the contrary, sustained playing together with the interlocutor was a typical feature of affiliatory behaviours, and well-segregated turn-taking was associated with disdain ... in speech, one does not signal affiliation by "talking" simultaneously over one's conversation partner ... In sum, the production and perception of these cues likely constituted a different cognitive domain than the interpretation of similar cues in speech. (106)

As I shall show, instead, people can talk simultaneously to the expression of affiliation and sociability, supporting a theoretical perspective that sees rhythm to convey social relations in both music and spoken language, that is, in *a single domain of embodied cognition of aesthetic and coordination cues.*

Prior to that—and within that same domain—consider dancing. I take the case of Western theatrical dance, on which I conducted extensive research, but ballroom dancing would be another good example. In dance, meaning is communicated via intentionally rhythmical movement. When dancing together, it is *coordinated* rhythmical movement at stake. In such circumstances, dancers exploit rhythm for mutual coordination, and for experiencing together the universe of meaning *integral* to the dance, the universe they simultaneously cocreate and inhabit (Bassetti, "Knowing"; *Genesi* 81–90, 140–44). Co-dancers produce meaning by interacting within the "universe of the dance" they are at once generating and experiencing. *That* is, actually, what is offered to the audience's perception and interpretation, what constitutes the "ongoing ... meaningful (interactional) context of the performance" ("Knowing" 107), and relations among dancers—of spatial, temporal, visual, and corporeal nature ("Knowing" 105–07; *Genesi* 81–90; Laermans 122)—are prominent material in that. Here the rhythm of action-in-interaction carries meaning, particularly social meaning.

Moving to the second "journey," it is well known the orchestra director employs rhythm in gesturing to convey rhythm in (co)playing, as the script is not enough. Dancers employ rhythmic talk when sharing with one another the kin/aesthetic details of movement, particularly for conveying the "dynamic" of dancing. When oriented to the conceptualization of rhythmical moving, some features of dancing can be verbalized (the "story" behind choreography, if any; "what" and "when" to perform, such as in "arabesque is on three" or notated choreography), but other kin/aesthetic *qualia*—"how" to perform, the quintessential core of expressive dance—require also paraverbal and nonverbal means such as prosody and rhythm to be made intersubjectively accessible (Bassetti, "Performative" 196; *Corpo*).[1] When a piece of dance is to be transmitted to performers, verbal and embodied prompting along-with dancers' enactment feature as prominently as demonstrating. Prompt talk is a highly rhythmical production that, as I illustrated elsewhere (Bassetti, "Performative" 183–84; Bassetti and Bottazzi 464–65), consists of mixing names (e.g., "plié," in Transcript 1[2]) and indexical expressions ("up," "stretch") for movements, with beat numbers (i.e., counting) and nonsemantic vocalizations ("whaaam").

Transcript 1
sei sette >e ↑plié< e un (.) e su (.) e >bum bum ta< plié e stendi e wha::m
six seven >and ↑plié< and one (.) and up (.) and >bum bum ta< plié and stretch and wha::m

The rhythmic dynamic of speech mimics that of dancing and can also convey other-than-rhythmical qualities of movement (e.g., "bum bum ta," produced in acceleration with respect to surrounding speech (> … <), speaks of a sharp, alongside fast way of moving). Noticeably, the higher the level of expertise, which also means the more artistic and expressive the action (rather than a technical exercise), the less semantically meaningful the talk, as in the following:

1 The local co-development of movement and discursive vocabularies, characteristic of choreographic cocreation in contemporary dance (Laermans 377), is another means to the same aim.
2 For notation signs in this and the following transcripts, cf. Jefferson ("Glossary"). English rendering is provided below the original Italian.

Transcript 2
Ia:: pa:: stra pa (0.1) an' ta (.) an' ta (.) an' tim an' ba.

What speech does here is completely besides its content; it approaches the "order of reality" of dancing. This is allowed by rhythm, here a tool for addressing aesthetic aspects of the performed.

Rhythm of Interaction

Prosody and Conversational Rhythms: Co-Creating a Local Social Order

Conversation, or talk-in-interaction, is an activity humans do together, on an everyday basis, that is characterized by the use of rhythm as one among the meaning-making tools available, and it is rhythmic in itself, as turn-taking could be thought of as (part of) the rhythm of conversation. In a book representing her lifetime work, neuroscientist Nina Kraus acknowledges how "*rhythm is a necessary ingredient of linguistic communication*" (111) and "[i]n all languages there is a definite rhythmic aspect to spoken language" (117). Within the social sciences, interactional linguistics as well as ethnomethodology and conversation analysis have studied poetics and prosody in naturally occurring talk, with incursions in the realm of acoustics and rhythm (e.g., Pelose; Couper-Kuhlen and Selting; Jefferson, "Poetics"; Auer et al.; Bassetti and Liberman, "Orderliness," "Making"). As "proto-conversation," "parentese," or "babytalk" has rhythm as a crucial component (Bateson; Gratier and Trevarthen). Analyzing mother–infant talk, Miall and Dissanayake showed its rhythmic shaping drives the baby's attention. They "hypothesize that the ability to respond to poetic features of [spoken] language is present as early as the first few weeks of life and that this ability attunes cognitive and affective capacities in ways that provide a foundation for the skills at work in later aesthetic production and response" (377).

The aesthetic *qualia* of action-in-interaction are indeed relevant, and particularly with respect to affect, sociality, and relations. A study by Floyd and Ray explored the role of three acoustic features of speech—namely, fundamental frequency (low/high pitch), the degree of its variation, and intensity (quieter to louder voice)—on perceived affection among interactants previously unknown to each other. They found pitch variance, pointing to rhythm, to influence both receivers' and observers' perception of the speaker's affection toward the receiver, whereas intensity was irrelevant and frequency per se relevant to observers only. One could say we reserve monochord talking to enemies and despicable people, it is with and for them that we do not bother to produce aesthetically pleasurable rhythmic talk (similarly, we tend to not mimic them, cf. Hess). More importantly, the study shows that *the aesthetics of performed language holds relational meaning.*

Research I conducted with Kenneth Liberman in Italy showed other aspects of rhythm alongside pitch variation to characterize sociable interaction, beyond dyadic conversation and speaker–receiver conceptualizations. Specifically, co-creating rhythm emerged as a way both to enjoy and display (produce and reproduce) sociable relations among participants, as rhythm "materializes in talk the harmony of their relationships" (Bassetti and Liberman, "Making" 101). "On sociable occasions, Italians concert their multi-party simultaneous speech precisely to constitute, animate and maximize the sociability of such occasions" (95). The aim is to encourage multiparty vocal participation. Several techniques are employed to achieve such an objective and to enjoy its developing results. One is repetition, of both one's and others' words as much as the acoustic and rhythmic features of talk, that is, "matching each other in volume, pitch, style of vowel-elongation and the prosodic contours of phrases in general" (102). Recurrence, together with variance, or difference, lies at the core of rhythm (Deleuze).

Notice the interlocked echoes across participants in Transcript 3 (Figure 6.1): S gets repeated by M ("Non lo sanno ancora," lines 1–2), who adds a vocal elongation which starts to set a rhythm and gets immediately employed by L (3–4, cf. further); lexically, M's talk is repeated by B ("è all'ospedale," lines 5, 9), who in turns finds his production picked up by C ("Cioè," 3, 9), whose "non è deeetto" (6) is repeated, with a descending rather than ascending tone (variation), by M (8), and whose "su internet" (11) in reused by L (15).

Transcript 3[3]

L's production in lines 3–4 not only reproduces vowel elongation firstly employed by M, but, with "intoller ↑ aanza alla g ↓ aaanja," also sets a clearer rhythm, allowing M to provide counterpoint to C's "non è deeetto" precisely half a bar later (Figure 6.2, arrows mark locations of "de"s, in the middle and at the

[3] Cf. Bassetti and Liberman, "Making" 101.

```
S  Non lo sanno ancora.
M  E::h //non lo sanno anco:ra >è all'ospedale<
L        //Cioè pensa che sfi:ga! C'ha un'intoller↑a:nza alla
   g↓a::nja*
B               *Ma è all'ospedale perché c'ha ripr[[ova:to? Cioè sapeva=
C                                                   [[No: non è de↑::tto.
B   =di ess°er{{e° (0.3) e g'he riprové?
M                  {{Non è de↓:tt:o.
C  Cio//è (.) è un'ipotesi di Giorgio che ha letto robe=
M     //Ma secondo me—
C  =su in:te[[rnet
B
V            [[A:h v{{a beh allo::ra
B                   {{A//:::::h  A::::::h
A                      //A:[[ah beh allo:ra
L                          [[A:::::::::h (.) s'è cura:to su internet
```

```
S  They still don't kno:w.
M  E::h t//hey still don't kno:w >he's at the hospital<
L        //I mean think what a bumme:r!  He has an intoler↑a:nce to
   g↓a::nja*
B               *But he's at the hospital as he tr[[ie:d again? I mean he=
C                                                 [[No: it's not s↑u::re
B   =knew he w°a{{s° (0.3) and he tried again?
M                  {{It's not s↓::rre
C  I me//an (.) it's a hypothesis by George who read stuff=
M     //But I think—
C  =on:li[[ine
V        [[A:h w{{ell the::n
B               {{A//:::::h  A::::::h
A                  //A:[[::h well the::n
L                      [[A:::::::::h (.) he cure:d himself online
```

Figure 6.1 "Transcript 3." Image generated by the author.

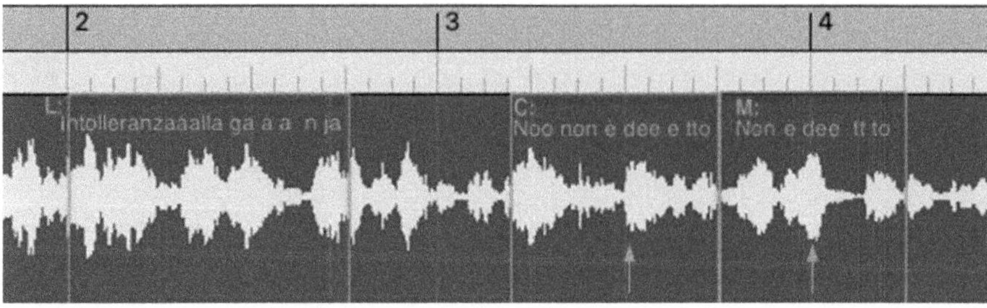

Figure 6.2 Waveforms of Transcript 3, lines 3–8 (125 BPM), metronome at the top of the figure. Image generated by the author.

end of bar #3). This eventually leads to a chorus of "Aaaaah," in a sort of canon (lines 12–15). It has been proposed (Cowley 560–61) that such matching "is particularly marked in languages …, as is the case for Italian, where it has been stylized by musical traditions."[4]

Another technique for setting and sustaining concerted rhythmic talk is based on vocal gestures, such as "Aaaaah" or "E↑ooooe↓eeee," as in the following example, which further demonstrates the role of self- and hetero-repetition ("una palla di pelo," lines 2, 4–5, 6).

Transcript 4 (Figure 6.3)[5]

"What C is doing in line 1 and line 8 is to *instruct* A and then L how to perform the song together … These duets work well in fulfilling the principal aim … which is to enhance the excitement and therefore the enjoyment of the parties." Indeed, for the rest of the interactional sequence, the prosodic contour "practically became an anthem that was iconic for the congenial relations they had achieved" (Bassetti and Liberman, "Making" 103). Enjoyment can also be spotted in collective laughing after rhythmic matching is accomplished:

Transcript 5 (Figure 6.4)

In sum, *the aesthetics of language coproduction holds social, relational meaning.* Even without considering cultural variation, other interactional formats than the convivial one here analyzed—think to a political debate, or the incipient moments of a gang fight (Collins, *Violence*)—manifest different rhythmical features in turn-taking (e.g., interrupting rather than singing along, competing

[4] Bryant distinguishes local and global prosody, where the latter "can also turn musical, with features like vowel prolongation (e.g., fixed pitch), rhythmic elements, and other melodic aspects" (203).

[5] Cf. Bassetti and Liberman, "Making" 102.

```
C  E↑o:::e↓::/:::
L              //Ma! (.) >E' una palla di pe[[lo!<
A                                                [[P↑o::ve::/ri↓:::no
M                                                                //E' una palla di
   pelo (.) eh si [[ce l'avev-
L                      [[>Una pa//lla di pelo<
C                                    //No:: due! Ne ha fatto uno anche là molto
   più gr↑o::ss[[o
L                    [[E↑o:::e↓::: (0.2) ha sboccaria::to

C  E↑o:::e↓::/:::
L              //But! (.) >It's a ball of f[[ur!<
A                                              [[P↑o::o::r //thi↓:::ng
M                                                                //It's a ball of
   fur (.) eh yes [[it had-
L                      [[>A ba//ll of fur<
C                                  //No:: two! It did one also there much
   b↑i::gge[[r
L                [[E↑o:::e↓::: (0.2) it barfe::d
```

Figure 6.3 "Transcript 4." Image generated by the author.

for turns at talk), and an overall interactional aesthetics, which, however, *equally* provide the occasion with relational meaning (e.g., antagonism). For instance, when conflict-based interaction is the operative format between individuals or groups,

> confrontations begin with staring at each other's face … arguing, making insults and gestures … mirroring each other's moves. In such arguments, people repeat the same phrase over and over again—it is useless to try to state a complex argument, and since both are talking at the same time, it is really a contest over getting the next conversational turn; a conflict on the paralinguistic level of speech acts, not in the contents of the speech itself. (Ferret and Collins 11)

In Aucouturier and Canonne, simultaneous playing by duetting musicians (improvising on instructions) was similarly found maximum both in their

```
W  La società dello zolfo
P  Okey (.) è una ro:ba alche:mica:: Cro::wle[[y: quell//e stronzate lì: Ah okey
W                                            [[Filosofi//a
F                                                      //Filosofia
N                                                      //Allo::ra >a quanto pare
   non fanno fiammiferi:< quin((laught[[er))di temo di sì (.) ma sarebbe mo::lto
P                                     [[((laughter))
N  meglio //(0.2) O::::h quella volta che ci siamo ritrova:ti quando lui studiava
F         //Sono (- - -) con gli ami-
N  sciamane[[simo:: tibet//↑A:::n↑o:
W          [[sim-        //beta::n↑o:
   ((communal laughter))
```

```
W  The society of solfur
P  Okey (.) is it alche:mica::l stu:ff Cro::wle[[y: thos//e bullshit like: Ah okey
W                                              [[Philoso//phy
F                                                       //Philosophy
N                                                       //So:: >apparently they
   don't do matche:s< hen((laught[[er))ce I'm afraid so(.) but it would be mu::ch
P                                [[((laughter))
N  better //(0.2) O::::h that time we met aga:in when he was studying
F         //I'm (- - -) with the fri-
N  Tibe[[ta::n sciam//↑A:::n↑i:sm tibet//↑A:::n↑o:
W      [[tan-       //ama::n↑i:sm
   ((communal laughter))
```

Figure 6.4 "Transcript 5." Image generated by the author.

"caring" (supporting) and "domineering" (monopolizing) categories of interaction. Specifically,

> domineering interactions were characterized by long overlaps, encoder-to-decoder causality and no occurrence of harmonic fusion; insolent and disdainful interactions were associated with short overlaps, no decoder-to-encoder causality, and harmonic contrast; caring, with long overlaps, decoder-to-encoder causality, and no occurrences of harmonic contrast. (Aucouturier and Canonne 101)

In short, there is something we hear in interaction—its material being music or talk—to which we attach relational meaning. The paralinguistic features of talk, rather than its semantic content, and rhythm of turn-taking seem to be what we use to characterize as such both conflictual and convivial occasions, and this characterization is a coordinated—either collaborative or competitive—endeavor in both cases. Conversation, and social interaction more generally is a rhythmic phenomenon we simultaneously create and experience together. In doing so, we continuously establish social order by defining and displaying the kind of social occasion at stake and the kind of relationships holding among participants, drawing also from aesthetic material to such an aim.

Interactional Aesthetics and Power, and How Rhythm Serves Charisma

Social relationships are also always power relations. In its most basic understanding, this translates into symmetrical versus asymmetrical relations. Another distinction is useful: the kind of power at stake. Drawing from Weber's seminal work and elaborating a micro-level, situated perspective on it (Bassetti and Bottazzi; Collins, *Charisma*; Joosse), we can distinguish between traditional, rational-legal, and charismatic authority. Whereas the first two kinds lie in formal roles ascribed to individuals, who enter (some) interactions with such a role and the power it conveys, although roles remain to be performed in interaction (cf. Goffman), charisma is based solely on performance and always open to assessment and revision (Weber §10, II). Charisma is a form of power that needs to be demonstrated any time, precisely in doing interaction. It is a situated kind of power—although repeated interaction can lead somebody to be deemed charismatic across interactional instances (which is how charisma has been read in most of the scientific literature, sociological or otherwise). The concept itself of charismatic leadership illustrates how charisma is an interactional accomplishment, when present, also for those holding formal power positions (e.g., managers, political leaders). Finally, charismatic power is marked by followers, rather than subjects. And following is an action rather than a status, something we do. At the micro-level of situated interaction, followers are *seduced* and *moved* to action by the leader, in whose capable hands they put themselves—charisma is persuasive.

In social interaction, power can be expressed, exercised, and acquired by different means. First, it can be claimed and/or "simply" exhibited. In exhibiting, turn-taking organization and the paralinguistic and acoustic features of collective talk, which build to the rhythm *of* interactions, are particularly relevant. These aspects both emerge out of underling power relations and social roles (e.g., students in class raise their hand to ask for a turn at talk whereas the professor does not) and make such relations and roles manifest in interaction, thereby continuously reproducing them—although conversation is also where existing power relationships can be contested and where power can be seized (or emerge). Participants co-create a rhythm and aesthetics of the interaction thereby bestowing onto the situation certain relational meaning, a social order.

Such co-creation, however, can be more or less symmetrical in itself. This is orthogonal to the power relationships at stake, opening up to power acquisition in interaction—that is, leader–follower dynamics do not only apply to situations characterized by asymmetric social roles, like the students–professor example, or employees–manager meetings, but might also emerge among peers; further, they can be absent or "masked" (cf. further) both among peers and with power inequalities holding. However, the a/symmetry of interactional co-production also affects extant relations. In the mentioned analysis of adult–newborn interaction, which entailed several co-present adults (Bassetti and Bottazzi 465–71), the adult rhythm-maker gains a special status—for the baby, whose attention and engagement are captured, but also and consequently for the whole group of interactants, who take an "audience" position, with a couple of them aiding the leading rhythm-maker by holding the baby in full silence (see Figure 6.5). This is how rhythm can be conducive to situated power, at the crossroads between nonconceptual and conceptual sense-making. The conceptual dimension preeminent in social contexts and roles "influences, but also is influenced by, the nonconceptual dimension of rhythm in situated embodied interaction" (Bassetti and Bottazzi 472).

In group conversation, the issue of guiding interaction, which both expresses and seizes power, can amount to emerging as the informal leader among peers, like in the above case, or demonstrating charismatic qualities when holding a formal position. On friendly occasions, peers jointly create a common rhythm, or one could say they continuously alternate in leading others to a common rhythm (as in Transcript 4, with C instructing A and L), which in turn celebrates and reinforces the horizontality and affiliativeness of their relationships. In competitive and confrontational interaction, leader–follower dynamics are basically the contested material at hand, and dominance of the "emotional attention space" (Collins, "Micro and Macro" 16) via rhythm is key. Liberman ("Logic" 394) offers an example of deceptive use of rhythm in his ethnomethodological analysis of Tibetan philosophical debating, featuring a defendant and a challenger.[6] Concerning conflictual interaction, as Collins

[6] On the role of interpersonal synchrony in deception see Duran and Fusaroli.

Figure 6.5 Snapshot from video-recordings of multiparty adult–newborn interaction. Image generated by the author.

notices based on a vast and variegated array of empirical data, it is rarely the case that confrontation escalates into physical violence. This is because "one is struggling not just against the other person, but against the tendency within one's own nervous and endocrinological systems to become entrained in the rhythms and emotions of the other" (Collins, "Micro-Sociology" 569). As mentioned previously, "[i]n ordinary non-conflictual interaction, there is a tendency … to be inducted into a common emotional mood, a shared definition of the situation" (Collins, "Micro and Macro" 16), "a tendency for persons who establish a common focus of attention to become mutually entrained in a common rhythm of speech and bodily movements," a "solidarity-producing tendency" (Collins, "Micro-Sociology" 569). In violence-threatening interaction this makes participants "subjected to deeply disturbing tension" (569). Often, this tension blocks violence to start and, after a short while, given its unpleasantness, interaction ends. However, when the tension gets overcome, violence develops its own rhythm, which is similarly difficult to resist (575), resulting, for example, in "overkilling":

> it is after one side has become emotionally weak, emotionally dominated in the interaction, that violence generally does its damage … as long as individuals (or groups) confront one another at cross purposes, they are … running contrary to the physiologically ingrained tendency to fall into a common rhythm with each other. Passivity on one side allows the other side to impose their own rhythm upon the interaction … the victim falls into participation with the rhythm of violence being perpertrated upon him or herself. (Collins, "Micro-Sociology" 571)

Rhythm can be persuasive, and when following, we may end up into suffering as much as into pleasure. This illustrates the "dark" potential of rhythm. Once rhythmic dominance has been secured, moreover, what were before conflicting peers become an aggressor and a victim, a winner and a loser. Power relations change at the conceptual level. Hip hop battles or ultras groups at the stadium ritually resemble this dynamic.

A recent study by Farrera and colleagues ("Local"), unconcerned with power and yet to be peer reviewed, explored the degree and directionality of the propagation of rhythmic information in naturally occurring small group conversation (work meetings among peers), also considering turn-taking dynamics. They found two main strategies of rhythmic coordination in group conversation, building to different emerging collective rhythms (cf. also Farrera and Ramos-Fernandez, "Collective"). I offer an interpretation of their results which draws on EMCA and partially departs from the authors' reading of deviant cases (*II.i*, *I.i* below). When turns at talk are equally distributed and there is variety in turn-taking patterns (who speaks after whom, i.e., reversely, who elicits whose talk or allocate a turn)—a situation similar to the sociable occasions Liberman and I explored in Italy[7]—the propagation of rhythmic information among participants is high and flows evenly among them. The emerging "global" rhythm (*I*) is less structured and holds a faster tempo, likely more engaging. Instead, when a "conversational hub" can be identified—somebody who talks more and after whom others intervene in greater proportion than after other members, which could also mean the hub is the main turn allocator—rhythmic propagation is smaller and flows mostly from the hub to the others. In sum, leader–follower dynamics are at play. This produces more structured collective rhythms with slower tempo (*II*). The

[7] "[T]he sociality of these conversations emphasizes a social context that is decentered" (Bassetti and Liberman, "Orderliness" 2).

same outcome emerges when somebody serves as conversational hub but (still smaller) rhythmic propagation is a collective effort (*II.i*).

Interestingly, in one instance of the dataset, one participant talks more than the others and guides rhythmic circulation, but there is variety of turn-taking patterns, that is, co-participants do not seem to feel "summoned" by his/her talk more than others' speech and a primary turn-allocator cannot be spotted. This generates a faster and less-organized global rhythm (*I.i*). To me, this is leading by setting the rhythm (also via talking more than the others, one needs a stand to instruct an ensemble), but in a way that goes unnoticed and creates a platform for anyone to join at any one time (although joining in a way that rhythmically aligns with the situated leader). The resulting "chaotic" interaction is a function of such a "masked" or "withheld" leadership. This, I believe, is one of the ways charismatic power manifests. For it has to be both beneficial (e.g., aiding group coordination) and engaging for the followers, so that "the real test is whether you can move people emotionally, get them focused, and make things happen" (Collins, *Charisma* 9).

I therefore see more than two strategies of rhythmic coordination, but two main outcomes, that is, collective rhythms with different qualities, different aesthetics, representing also relational meaning. A symmetric coproduction in conversation can appear chaotic and is achieved chorally, but the chorus can exploit none or alternating (*I*), or "withheld" (*I.i*) directors. Both ways end up materializing in interaction the horizontality of the relationships at hand. On the other hand, centralized co-productions appear more ordered and are accomplished through leader–follower dynamics, whether participants rhythmically align mostly with the leader, like solo players in an orchestra (*II*), or also with each other, like the group of violins (*II.i*). In both cases, the interaction displays asymmetry although, formally speaking, among peers. This is one way to gain power informally: by guiding others into an interaction that ends up representing an asymmetry. Whereas (*I*) is characterized by sustained coordination, (*II*) sees participants moving in and out coordination states. The former situation, with also its faster rhythm, is more engaging and provides "emotional energy" (Collins, *Interaction*). Further, although apparently counterintuitively, mutual listening is higher in "chaotic," decentralized conversations (Bassetti and Liberman, "Making" 110–11, "Orderliness" 14–19).

Now, if we consider lead–follow dynamics, we see three of four strategies employ them but one does so in a "withheld" way, achieving emotional energizing, which is characteristic of charismatic leading (Collins, *Charisma*).

Moreover, this strategy displays and establishes horizontal relations, a symmetric local social order (although in the presence of what we might call a "*primus* hidden *inter pares*"), which characterizes charismatic groups. Overall, a charismatic "community (*Gemeinde*) … is based on an emotional form of communal relationship" (Weber §10.III). The fact that two of the strategies employing leading (*II.i*, *I.i*) were enacted by the same participant demonstrates how charismatic leadership is, first and foremost, a situated interactional accomplishment rather than a personal trait.

Guiding an interaction rhythmically, rather than topically and/or in its turn-allocation structure, is leading at a level that remains behind and beyond the conceptual dimension. That is, rhythm seems to allow to gain power informally in interaction at the level of seduction, rather than—or together with—in terms of content or speaking rights management. Rhythm serves charismatic leading specifically: charisma persuades others to act by seduction, beyond their conscious awareness, and rhythm is key in such an endeavor, both in terms of the aesthetic dimension of the leader's performance (e.g., D'Errico et al.) and as an emerging collective achievement resulting in an engaging, energizing, pleasurable interaction.

Conclusion

Relational and hence power-related meanings are produced and interpreted in social interaction—in music, dance, and everyday conversation of convivial, conflictual, or work-oriented nature. Rhythm seems particularly suited to express relational meaning, hence to deal with power in interaction. Symmetrical affiliative relationships can be displayed by sustained rhythmic alignment and, particularly when the occasion is convivial, by extended simultaneous talking, paralleling harmonic consonance and simultaneous playing in music, or motor synchrony in dancing or walking together. One "We" is identifiable; everyone is emotionally energized. Symmetric but conflictual relations are expressed through overlapping talk but as a fight over setting the rhythm, hence gaining emotional dominance, rather than a choral coproduction. Two sides are evident (two "We" in group fights). On the contrary, asymmetrical power relations are either reproduced or emerge in interaction by providing it with a more ordered aesthetics, involving slower rhythms and substantive leading–following, if not fully hierarchical dynamics. A "One" and some "Others"; none is thrilled. By using rhythm *in* interaction, we create the rhythm *of* interaction, building to its overall aesthetics. That is, rhythm is one of the tools we use to produce interactional aesthetics, able to carry relational meaning while affecting us at the embodied, affective level.

This makes rhythm also a means for managing power and power relations in interaction, particularly for doing so at the nonconceptual extreme of the continuum. This way, it serves charisma more than other forms of situated power, including when the latter is exhibited at a micro but more conceptual level by holding the reins of conversational turns, either based on explicit hierarchies or among peers. Notice that turn-taking/allocation issues can be made explicit in everyday discourse and conversation (e.g., "Do not interrupt," "You must let also the others speak"), whereas that is not the case for rhythmic leading/imposing (it can be made explicit, but that happens within expert rhythmical practices such as improvised music or dance). "Charisma is the product-in-being-produced of social interaction, setting the rhythm of the latter equals seizing that micro-power" (Bassetti and Bottazzi 473), and this happens implicitly. Accordingly, emotional energy, differently than emotions, is fully nonconceptual, and it is energizing, or arousal, and the production of a sense of togetherness, of we-ness—whether, for example, in a sad or happy mode—the charismatic person is capable of. That is the actual "proof" of charisma. This is a peculiar form of leading, where a "We" is identifiable, although hiding a *primus inter pares* within.

Overall, rhythm can be a means for gaining power, either by charismatic seduction, arousing energy in others, or by "winning" emotional dominance over them in a conflict. On the other hand, rhythm can be a way to exhibit the absence of power imbalances. When power is managed this way, that is, nonconceptually, in situated embodied interaction, it in turn generates an effect at the conceptual level—the appearance of a leader with their group, of a "winner" side in confrontation, or of a close-knit group of peers—with consequences for future interaction. "The strength of rhythm is not simply due to its nonconceptual associative, action-laden, and emotional character, but to its *social* nonconceptual associative, action-laden, and emotional character" (Bassetti and Bottazzi 473). Rhythm carries relational meaning and allows to produce interactional aesthetics capable of bridging the nonconceptual and conceptual layers of human communication and social order. Action-in-interaction is where this happens.

As often the case in life, the power of rhythm can be used to both positive ends, such as soothing others, creating intimacy, enjoying conviviality together or making art, and negative ones, like deceiving others, inducing them into passivity (Collins, "Micro-Sociology" 571) or "harmful obedience" (Gelfand et al.), until group torture (think, e.g., to Abu Ghraib facts during the US "war on terror") and overkilling. The better we understand how rhythm functions in interaction and, more generally, what is the import of aesthetics in interaction—rather than treating it as a separated domain—the better we could deal with "dark" outcomes.

Works Cited

Aucouturier, Jean-Jacques, and Cécile Canonne. "Musical Friends and Foes: The Social Cognition of Affiliation and Control in Improvised Interactions." *Cognition*, vol. 161, 2017, pp. 94–108.

Auer, Peter, Elizabeth Couper-Kuhlen, and Franz Müller. *Language in Time: The Rhythm and Tempo of Spoken Interaction*. Oxford UP, 1999.

Bassetti, Chiara. *Corpo, Apprendimento e Identità: Sé e Intersoggettività nella Danza*. OmbreCorte, 2021.

Bassetti, Chiara. *Genesi dell'Opera d'Arte: Fare Danza Assieme*. Mimesis, 2019.

Bassetti, Chiara. "The Knowing Body-in-Action in Performing Arts: Embodiment, Experiential Transformation, and Intersubjectivity." *Artistic Practices, Social Interactions and Cultural Dynamics*. Edited by Tasos Zembylas. Routledge, 2014, pp. 91–111.

Bassetti, Chiara. "Performative Teaching and Learning. On the Instruct-ability of Kin/aesthetic Properties." *Instructed and Instructive Actions: The Situated Production, Reproduction, and Subversion of Social Order*. Edited by M. Lynch and O. Lindwall. Routledge, 2023. https://doi.org/10.4324/9781003279235.

Bassetti, Chiara, and Emanuele Bottazzi. "The Power of Rhythm: From Dance Rehearsals to Adult–Newborn Interaction." *Etnografia e Ricerca Qualitativa*, vol. 8, no. 3, 2015, pp. 453–80.

Bassetti, Chiara, and Kenneth Liberman. "Making Talk Together: Simultaneity and Rhythm in Mundane Italian Conversation." *Language & Communication*, vol. 80, 2021, pp. 95–113.

Bassetti, Chiara, and Kenneth Liberman. "The Orderliness and Sociability of 'Talking Together': Portrait of a Conversational Jam Session." *Social Interaction: Video-Based Studies of Human Sociality*, vol. 4, no. 1, 2021.

Bateson, Mary C. "Mother–Infant Exchanges: The Epigenesis of Conversational Interaction." *Annals of the New York Academy of Sciences*, vol. 263, 1975, pp. 101–13.

Bress, Sophie. "The Art of Dancing Without Music." *Dance Magazine*, March 12, 2024, www.dancemagazine.com/dancing-without-music/#gsc.tab=0.

Bryant, Gregory. "Vocal Strategies in Verbal Irony." *The Cambridge Handbook of Irony and Thought*. Edited by Raymond W. Gibbs, Jr. and Herbert L. Colston. Cambridge UP, 2023, pp. 197–215.

Cirelli, Laura K., Kathleen M. Einarson, and Laurel J. Trainor. "Interpersonal Synchrony Increases Prosocial Behavior in Infants." *Developmental Science*, vol. 17, 2014, pp. 1003–011.

Collins, Randall. *Charisma: Micro-sociology of Power and Influence*. Routledge, 2020. https://doi.org/10.4324/9780429353314.

Collins, Randall. *Interaction Ritual Chains*. Princeton UP, 2004.

Collins, Randall. "Micro and Macro Causes of Violence." *International Journal of Conflict and Violence*, vol. 3, no. 1, 2009, pp. 9–22.

Collins, Randall. "The Micro-Sociology of Violence." *British Journal of Sociology*, vol. 60, no. 3, 2009, pp. 566–76.

Collins, Randall. *Violence: A Micro-Sociological Theory*. Princeton UP, 2008.

Couper-Kuhlen, Elizabeth, and Margret Selting, eds. *Prosody in Conversation: Interactional Studies*. Cambridge UP, 1996.

Cowley, Stephen J. "Of Timing, Turn-Taking, and Conversations." *Journal of Psycholinguistic Research*, vol. 27, no. 5, 1998, pp. 541–71.

Cummins, Fred. *The Ground from Which We Speak: Joint Speech and the Collective Subject*. Cambridge Scholars, 2018.

Deleuze, Gilles. *Différence et répétition*. PU de France, 1968.

D'Errico, Francesca, et al. "The Perception of Charisma from Voice: A Cross-Cultural Study." *Humaine Association Conference on Affective Computing and Intelligent Interaction*, Geneva, Switzerland, 2013, pp. 552–57.

Duran, Nicholas D., and Riccardo Fusaroli. "Conversing with a devil's advocate: Interpersonal coordination in deception and disagreement." *PLoS One*, vol. 12, no. 6, 2017, e0178140.

Durkheim, Émile. *The Elementary Forms of Religious Life*. Free Press, 1965.

Farrera, Arodi, and Gabriel Ramos-Fernández. "Collective Rhythm as an Emergent Property During Human Social Coordination." *Frontiers in Psychology*, vol. 12, 2022, 772262.

Farrera, Arodi, Caleb Rascon, and Gabriel Ramos-Fernandez. "Local and Global Rhythmic Dynamics in Small-Group Conversations." *Research Square*, 2024. Online. https://doi.org/10.21203/rs.3.rs-3888842/v1

Ferret, Jérôme, and Randall Collins. "On the Internal Dynamics of the Conflict/ Violence Process: A Discussion with Randall Collins." *The American Sociologist* vol. 49, 2018, pp. 5–15. https://doi.org/10.1007/s12108-017-9346-0.

Floyd, Kory, and George B. Ray. "Vocalic Predictors of Perceived Affection in Initial Interactions." *Western Journal of Communication*, vol. 67, no. 1, 2003, pp. 56–73.

Gelfand, Michele J., et al. "The Cultural Evolutionary Trade-Off of Ritualistic Synchrony." *Philosophical Transactions of the Royal Society B*, vol. 375, 2020, e20190432.

Gill, Satinder P. "Rhythmic Synchrony and Mediated Interaction: Towards a Framework of Rhythm in Embodied Interaction." *AI & Society*, vol. 27, no. 1, 2012, pp. 111–27.

Goffman, Erving. *Encounters: Two Studies in the Sociology of Interaction*. Bobbs-Merrill, 1961.

Golland, Yulia, Yossi Arzouan and Nava Levit-Binnun. "The Mere Co-Presence: Synchronization of Autonomic Signals and Emotional Responses Across Co-Present Individuals Not Engaged in Direct Interaction." *PLoS One*, vol. 10, no. 5, January 2015, pp. e0125804.

Gordon, Ilanit, et al. "Physiological and Behavioral Synchrony Predict Group Cohesion and Performance." *Scientific Reports*, vol. 10, no. 1, 2020, 8484.

Gratier, Maya, and Colwy Trevarthen. "Musical Narrative and Motives for Culture in Mother–Infant Vocal Interaction." *Journal of Consciousness Studies*, vol. 15, nos. 10–11, 2008, pp. 122–58.

Hess, Ursula. "Who to Whom and Why: The Social Nature of Emotional Mimicry." *Psychophysiology*, vol. 58, no. 1, 2021, e13675.

Hove, Michael J., and Jane L. Risen. "It's All in the Timing: Interpersonal Synchrony Increases Affiliation." *Social Cognition*, vol. 27, no. 6, 2009, pp. 949–61.

Jefferson, Gail. "Glossary of Transcript Symbols with an Introduction." *Conversation Analysis: Studies from the First Generation*. Edited by G. H. Lerner. John Benjamins, 1984, pp. 13–31.

Jefferson, Gail. "On the Poetics of Ordinary Talk." *Text and Performance Quarterly*, vol. 16, no. 1, 1996, pp. 1–61.

Joosse, Paul. "Becoming a God: Max Weber and the Social Construction of Charisma." *Journal of Classical Sociology*, vol. 14, no. 3, 2014, pp. 266–83.

King, Anthony, and Mark de Rond. "Boat Race: Rhythm and the Possibility of Collective Performance." *The British Journal of Sociology*, vol. 62, no. 4, 2011, pp. 565–85.

Koch, Sabine C. "Rhythm Is It: Effects of Dynamic Body Feedback on Affect and Attitudes." *Frontiers in Psychology*, vol. 5, 2014, 537.

Kraus, Nina. *Of Sound Mind: How Our Brain Constructs a Meaningful Soundscape*. MIT P, 2022.

Laermans, Rudi. *Moving Together: Theorizing and Making Contemporary Dance*. Valiz, 2015.

Lakens, Daniël, and Mariëlle Stel. "If They Move in Sync, They Must Feel in Sync: Movement Synchrony Leads to Attributions of Rapport and Entitativity." *Social Cognition*, vol. 29, no. 1, 2011, pp. 1–14.

Lang, Martin, et al. "Lost in the Rhythm: Effects of Rhythm on Subsequent Interpersonal Coordination." *Cognitive Science*, vol. 40, no. 7, September 2016, pp. 1–19.

Liberman, Kenneth. "Objectivation Practices." *Social Interaction: Video-Based Studies of Human Sociality*, vol. 1, no. 2, 2018.

Liberman, Kenneth. "The Logic Is Made to Dance: Rhythm in Tibetan Debating." *Etnografia e Ricerca Qualitativa*, vol. 8, no. 3, 2015, pp. 383–97.

McNeill, William H. *Keeping Together in Time: Dance and Drill in Human History*. Harvard UP, 1995.

McPhail, Clark. *The Myth of the Madding Crowd*. de Gruyter, 1991.

Miall, David S., and Ellen Dissanayake. "The Poetics of Babytalk." *Human Nature*, vol. 14, 2003, pp. 337–64.

Mogan, Reeneta, Ronald Fischer, and Joseph A. Bulbulia. "To Be in Synchrony or Not? A Meta-Analysis of Synchrony's Effects on Behavior, Perception, Cognition, and Affect." *Journal of Experimental Social Psychology*, vol. 72, 2017, pp. 13–20.

Oullier, Olivier, et al. "Social Coordination Dynamics: Measuring Human Bonding." *Social Neuroscience*, vol. 3, 2008, pp. 178–92.

Palumbo, Richard V., et al. "Interpersonal Autonomic Physiology: A Systematic Review of the Literature." *Personality and Social Psychology Review*, vol. 21, no. 2, 2017, pp. 99–141.

Pelose, Gina C. "The Functions of Behavioral Synchrony and Speech Rhythm in Conversation." *Research on Language and Social Interaction*, vol. 20, nos. 1–4, 1987, pp. 171–220.

Safron, Adam. "What Is Orgasm? A Model of Sexual Trance and Climax via Rhythmic Entrainment." *Socioaffective Neuroscience & Psychology*, vol. 6, no. 1, 2016.

Savage, Patrick E., et al. "Music as a Coevolved System for Social Bonding." *Behavioral and Brain Sciences*, vol. 44, 2020, e59.

Schutz, Alfred. "Making Music Together: A Study in Social Relationship." *Collected Papers*, vol. 2. Martinus Nijhoff, 1971, pp. 159–78.

Vicaria, Imelda M., and Lucy Dickens. "Meta-Analyses of the Intra- and Interpersonal Outcomes of Interpersonal Coordination." *Journal of Nonverbal Behavior*, vol. 40, 2016, pp. 335–61.

Vuoskoski, Jonna K., et al. "Feeling Moved by Music: Investigating Continuous Ratings and Acoustic Correlates." *PLoS One*, vol. 17, no. 1, 2022, e0261151.

Weber, Max. "*Die Typen der Herrschaft*." *Wirtschaft und Gesellschaft*, 1925, vol. 1, pt. 1, ch. 3.

7

Routines of Creativity: Labor, Art, and the WPA

Caroline Levine

There are few aspects of modern life that artists and theorists have hated more than routine. The deadening monotony of factory work, the unthinking acceptance of custom, the recycling of familiar images and ideas. Mechanization. Habit. Convention. Stereotype. Cliché. Critics since the Romantics have insisted on the value of surprise and unpredictability. As Ralph Waldo Emerson put it in 1841, "People wish to be settled; only as far as they are unsettled is there any hope for them" (413).

Typically, art is hailed as the best antidote to the anesthetizing power of routine. According to Percy Shelley, for example, poetry is valuable because it "strips the veil of familiarity from the world" (139). Viktor Shklovsky argues that it is art's defamiliarizing force that jolts us out of the deadening routines of modernity. The painter Jean Dubuffet wrote that art's job is "to smash all the crust of routine, to crack open the shell of policed and socialized man, to unblock the channels through which the voices of his internal wild man can express themselves" (qtd. Maclagan 49). John Dewey agrees: "Art throws off the covers that hide the expressiveness of experienced things; it quickens us from the slackness of routine" (104). This tradition continues in our own moment with thinkers such as Derek Attridge, who argues for literature as interrupting a "permanent settling of norms and habits" (137).

If routine dulls experience, rhythm, critics suggest, does the opposite. Associated with the aesthetic pleasures of music, poetry, and dance, rhythm has long been considered expressive, liberatory, linked to the very pulse of life: the Greek *iambos* means heartbeat, and in Sanskrit *pāda* means footstep. Even the Modernists who committed themselves to breaking from conventions embraced rhythm as essential to art. From Fernando Pessoa, who defines rhythm as "the intimate movement of the soul" (qtd. Barletta x) to Virginia Woolf, who imagines writing as rhythmic waves in the mind that precede words (qtd. Simpson 41), to Amiri Baraka's "holy nuance beating against itself" (199), a vast range of modern artists and thinkers have celebrated rhythm while denouncing routine. Disciplines in the university reinforce this distinction, as sociologists and historians study routine as part of the standardization of life by modern clocks and machines, while biologists and musicologists pay attention to rhythm.

This essay argues against the long-standing distinction between oppressive routine and liberating rhythm. Taking a formalist's perspective, I will argue that rhythms and routines have more in common than this dominant tradition of thought has acknowledged, and will make the case that routine enables artists in ways that have been both aesthetically and politically significant.

Where routine carries associations of endless sameness (from the French *route*, taking the same road again and again), rhythm (from the Greek *rhythmos*) suggests a more organic phenomenon, referring to beats, durations, and pulses. For over two centuries critics of industrialization have rejected routine in part for its rigid constraints on the organic body. As Friedrich Engels puts it: "the worker's activity is reduced to some paltry, purely mechanical manipulation, repeated minute after minute, unchanged year after year" (119). The same objections remain strong into our own time, as enforced routines organize work globally, including in sweatshops and electronics manufacture.

But on closer inspection, the distinction falls apart. As forms, rhythm and routine are difficult to distinguish; both entail repetitions occurring at measured intervals; both organize temporal experience according to predictable patterns. And although we might associate routines with the artificial or mechanical repetitions of modern life, we often refer to the rhythms of work and leisure and the rhythms of the city (Lefebvre 27). The ancient Greeks themselves defined rhythm as a patterned constraint on dynamic flow and movement—not altogether distinct from the kind of artificial imposition of order which today we might most closely associate with routine (Lynch 276). In practice, too, it can be hard to distinguish between artificial and organic patterns. Is falling asleep at night a rhythm or a routine? (Moss et al.). Is family mealtime a response to the rhythmic needs of the body, a basic drive to community, an effect of compulsory heterosexuality, an accommodate of modern laboring routines, or—as I would argue—all of those, tied up together?

The usual critique of routine is political. William James makes the argument that habit—which I take here as synonymous with routine[1]—keeps injustice going. In a famous passage, he writes that habit is "the enormous fly-wheel of society .... It alone is what keeps us all within the bounds of ordinance, and saves the children of fortune from the envious uprisings of the poor" (James 122). Yet, in a much less famous paragraph that immediately follows this one, James argues that a total absence of routine is actually worse:

> There is no more miserable human being than one in whom nothing is habitual but indecision, and for whom the lighting of every cigar, the drinking of every cup, the time of rising and going to bed every day, and the beginning of every bit of work, are subjects of express volitional deliberation. Full half the time of such a man goes to the deciding, or regretting, of matters which ought to be so ingrained in him as practically not to exist for his consciousness at all. (122)

Routine in small ordinary matters, James explains, releases precious mental energy for more interesting activity, including imagination and art.

The historical record confirms James's argument. Vast numbers of creative artists have relied on routine.[2] Percy Shelley "was up at six or seven, reading Plato, Sophocles, or Spinoza, with the accompaniment of a hunch of dry bread" (Trelawny 140–41). Gustave Flaubert kept his whole household to a strict daily regimen: perfect quiet before 10 a.m., then hair tonic, a pipe, and conversations with his mother. He wrote from 1 p.m. until 1 a.m. (Brown 293). Anthony Trollope, Wallace Stevens, Octavia Butler, and Toni Morrison wrote early in the morning before going to their day jobs. W. H. Auden timed breakfast, writing, shopping, and drinking to the minute, checking his watch constantly (Schott 53). Derek Walcott described "the early dark and the sunrise, and being up with the coffee and with whatever you're working on" as "a very ritualistic thing" (Hirsch 198). And here's Maya Angelou: "I keep a hotel room in which I do my work—a tiny, mean room with just a bed, and sometimes, if I can find it, a face basin. I try to get there around 7, and I work until 2 in the afternoon. ... It's lonely, and it's marvelous" (qtd. Tate 155).

[1] Habit, like rhythm, is usually considered biological and embodied, and set against the artifices of routine. But habit, like both rhythms and routines, is a form of repetitive organization, and it too can be socially enforced and constraining.

[2] I have learned about many of the following examples from Currey's *How Artists Work* and *Women at Work*.

Visual artists and composers, too, have kept to strict schedules to preserve their time for making art. Chris Ofili reports that he "arrives in his studio at 9 or 10 in the morning …. First, he tears a large sheet of paper, always the same size, into eight pieces, all about 6 by 9 inches. Then he loosens up with some pencil marks, 'nothing statements, which have no function'" (Kimmelman). Le Corbusier's routine, as his assistant described it, was "implacably regular": the day started at 6 a.m. with gymnastics and painting, a "kind of fine arts calisthenics." There was trouble for assistants who did not respect the clock. "The hour of 2:00 P.M., I soon learned, was holy. If you were a minute late you risked a reprimand" (Soltan ix, xiv). The composer Erik Satie took clockwork precision particularly literally: "Up at 7:18. From 10:23 to 11:47: inspiration. Lunch at 12:11. Leave the table by 12:14. … More inspiration: 3:12 to 4:07. Bed at 10:37. 'Once a week I wake up with a start at 3:19,' he wrote. That's on Tuesday" (qtd. Kimmelman).

One might assume that routines give shape to creativity only under modern capitalism, but there is evidence that artists and writers have relied on it for centuries. Working on *The Last Supper*, Leonardo da Vinci "would arrive early, climb up on to the scaffolding, and set to work. Sometimes he stayed there from dawn to sunset, never once laying down his brush, forgetting to eat and drink, painting without pause" (qtd. Pierucci 116–17). Anne Bradstreet, the seventeenth-century New England poet, found time to write at night after caring for her eight children (Currey, *Women at Work* 185). And Immanuel Kant's neighbors, famously, knew that it was 3:30 p.m. exactly when he left his house for his daily walk (Kuehn 14).

To be sure, not all artists have followed routines. Picasso had notoriously chaotic habits (Miller 96–7), and Walker Evans, constantly distracted, "would wake up, have breakfast, go back to bed, get up again" (Thompson 31). But I want to make the case that as ways of imposing order—as *forms*—routines are not only oppressive, they are also capacitating. Setting aside regular, protected blocks of time for work frees artists from having to make decisions about what each day will look like, reduces distractions, and maintains the momentum of ongoing creative labor. The result is that many artists oppose routinization as crushing in theory while relying on it in practice. Polish painter Helena Minginowicz puts it this way: "I generally dislike routine, but the kind that allows me to maintain a high quality of work is both special and necessary, so I fully embrace it" (Palma). Flaubert cast the relationship in more dialectical terms: "Be regular and orderly in your life," he advised, "so that you may be violent and original in your work" (280).

One might argue that the real problem with routine lies in the fact that it is imposed from without—demanded from workers by government and business. But I want to turn here to the New Deal Arts programs of the 1930s and 1940s to make the case that even imposed, routine forms of industrial labor can be crucially enabling for artists, and especially for those from marginalized groups.

In June 1935, US President Franklin Delano Roosevelt launched the Works Progress Administration to create jobs for the millions of Americans thrown out of work during the Depression. Men without specific skills were put to work building infrastructure—including sewers, bridges, dams, and airports—while a smaller number of women were employed in conventionally feminine jobs, such as caring for the elderly, canning food for schools, and sewing clothes for needy families (Swain). The government also created jobs for skilled workers in research, nursing, teaching, and typing.

Among the work relief programs were four federally funded arts projects, employing visual artists, actors and directors, musicians, and writers. The highest priority of these projects was, like the rest of the WPA, to get indigent people off the dole ("Hell," said director Harry Hopkins, "[artists have] got to eat just like other people" (qtd. Opdycke 73)). The second was to match the skills to the labor. As Aubrey Williams, Assistant Federal Relief Administrator put it, "We don't think a good musicians should be asked to turn second-rate laborer in order that a sewer may be laid rather than a concert be given for the pleasure of our people" (qtd. Taylor 245). And the third was to democratize the arts—to bring American culture to a broad range of audiences in every state. Artist George Biddle advised the President that the great Mexican muralists had succeeded in creating a powerful national art because the government had employed them "at plumbers' wages" (Biddle and Phillips 93). The result was that cultural workers, like everyone else on the WPA relief rolls, would be paid a basic wage—usually around $24/week, just enough to cover food and rent.

From the start, the press charged that these cultural projects were wasteful. The *Baltimore Sun* called the workers "sorry daubers, ham actors, spavined dancers, and radical scriveners luxuriating on the dole" (qtd. Opdycke 74). In response, WPA administrators tried to prove that their arts employees were just like other workers, putting in a full work week and engaged in regular production. But this proved an ongoing challenge. It is easy enough to pay workers when "you want to build a road, or build an airport," said Robert Asure, a WPA administrator: "but how do you put an artist on a payroll? Should he be paid by the hour, or by the yard, or who should supervise him, what constitutes production?" (Asure and Phillips 3).

The WPA experimented with a variety of ways to fit artistic labor to the model of industrial work (Hapke 3). The Federal Music Project, for example, insisted on rigid rehearsal and performance schedules, demanding eight performances every week (Bindas 8). But this did not result in a sufficiently strict timetable. In the first year of the project in New Orleans, musicians were required to rehearse together for three hours in the morning and then perform around the city at schools, baseball games, and retirement homes in the afternoon. This meant that their workday often ended early. When FMP administrators learned that some musicians were moonlighting in the evening, thus undercutting musicians not employed by the WPA, they responded by requiring a fully planned day of practice and shows. Musicians who missed roll call would be fired (Abate 11).

The Federal Art Project had an even harder time figuring out how to ensure that painters in home studios were putting in full-time hours. The Easel Division decided to require a finished 16″ × 20″ canvas every four weeks, or a 24″ × 36″ every six weeks (Solman 118). Local units also experimented with punch clocks and time sheets. In the first years of the project, the New York City Easel Division required painters to check into work at a Manhattan office every morning at 8 a.m. Jackson Pollock was often seen rushing to the office in his pajamas seconds before the clock struck 8 (Solomon 80). The project ended this requirement after artists in the outer boroughs complained that they were spending too much of their workdays getting back and forth to Manhattan. By the time painter Louis Bunce moved to New York in 1940, time checkers were visiting painters at their homes. "They would come around maybe not every day, but maybe every other day and see if you were working" (Bunce and Bestor 20). If artists wanted to make a sketch in some other part of the city, they were expected to notify the timekeepers exactly where they would be and when (Kainen 163). In some states, artists had to work in cubicles in tightly supervised large studios. Seattle painters, for example, "had to report every day to the Bailey-Gatzert School and work there, and … were checked up on for the amount of time they put in, even the amount of time they took off for lunch or to go to the washroom" (Chong and Bestor 4).

Printmakers and Poster Division workers were easier to supervise than painters and sculptors, because they were expected to show up at regular times to collaborative workshops (Pillen 15). Some were given assignments to make print runs of posters that would advertise other government-funded initiatives, such as national parks, smallpox vaccines, and WPA-funded band concerts. Poster units were a lot like small production lines.

Mural painters' labor was fairly visible in its own way, since the public could often see the workers painting the walls of town halls and post offices. For her mural in Coit Tower in San Francisco, Maxine Albro was as much a manager as an artist, supervising a whole team that included a mosaic-setter, a paint mixer, and two plasterers (Albro and Hall 3, 4, 6). Muralists often complained that WPA administrators were inclined to intrude at every step. Pollock so much disliked the teamwork required by mural painting that he transferred to the Easels Division as soon as he could (Solomon 80).

The Writers' Project typically modeled local units on newsrooms, sending workers out into the field to collect stories for the state guides and imposing regular deadlines for material. The Indiana office expected 1,200–2,000 words a week (Blakey 42). Jay du Von, who administered the Iowa Writers' Project, introduced time sheets (Du Von and Hewes). (He also noted that his people voluntarily worked extra hours at no pay.) But it was often frustrating to measure the labor hours of writers on the WPA payroll, because most worked alone on the road. Zora Neale Hurston, already a famous writer who was employed by the Florida Writers' Project, would disappear for long stretches. Stetson Kennedy, then one of her coworkers in Florida, would write Hurston a letter "to jog her up." In response, "we would receive a thick packet of fabulous folksongs, tales, and legends .... We did not care how, where, or when Zora had come by them—each and every one was priceless" (qtd. Boyd 314).

One site where art and industry came especially close together was the short-lived Design Laboratory in New York City, which taught free classes on Bauhaus principles and the "International Style," making ordinary objects such as chairs and cigarette holders. Instructors were at pains to distinguish designers from the stereotype of the individual artistic genius, stressing that design was "just one part of the manufacturing process," "in no more exalted a position than the tool and die maker in the factory" (Clark 45). The WPA cut off funds to this unit after four years because of its close ties to union organizing and communism.

Despite the WPA's best efforts, public charges of wastefulness and subversiveness continued, and arts administrators stepped up their surveillance. Washington instituted a Complaints Bureau to investigate charges of workers "loafing" on federal wages. In late 1936, New York "instituted loyalty oaths, paid armed guards to monitor demonstrations, and employed spies to identify, fire, and blacklist between twenty and forty thousand agitators and shirkers on the relief rolls" (Musher 175–76).

Many artists bitterly objected to the WPA's attempts to routinize their labor. Writers complained about "hack work" (Jackson 52). Painters argued with timekeepers, resenting their disciplinary surveillance (Clark 149–50). Albert C. Barnes, famous Philadelphia art collector, led a public campaign to remove the director of the Pennsylvania Federal Art Project, Mary Curran, because she insisted that artists had to work fixed hours in their studios every weekday, which prevented them from taking classes at the Barnes Foundation. This "un-American administration," Barnes's people insisted, created "a hindrance to a genuine art movement" (Meyers 208–09).

Humanities scholars might well agree. For many years, artists and critics have dismissed the visual works of the period as shoddy and unoriginal, often assuming that governments curb creative autonomy (Jeansonne 161). And some have argued that the WPA played a role in turning creative arts into advertising and commodities, paving the way for the consumer capitalism that would dominate the rest of the century (Fraser). Robert Motherwell, looking back on the Federal Arts Project, said "Nothing good came out of it, really nothing" (Shahn and Phillips 4).

But Motherwell was wrong. Numerous visual artists, writers, directors, actors, and musicians got their start thanks to the WPA. David Smith remembered that he, Pollock, and de Kooning "all came from the bond of the WPA." And it was good for their morale: "for the first time collectively we belonged somewhere" (Sylvester, 3). Painter Seymour Fogel said that the Federal Art Project "not only saved my life financially but launched my career as an artist" (Opdycke 90).

Artists of color, in particular, repeatedly traced their aesthetic successes to the routines of regular wages and blocks of time. Although Black centers of art and culture such as Harlem and Chicago's South Side were already dynamic hubs, WPA employment was still important in that it allowed a number of African American artists the time away from other paid work to dedicate themselves to art. Ralph Ellison is one example: "I could not have become a writer at the time I began had I not been able to earn my board and keep by doing research for the New York project." He argued that "the WPA provided an important surge to Afro-American cultural activity" generally, "allowing many Negroes to achieve their identities as artists" (204–05). Richard Wright told a similar story. Growing up Black in Mississippi, he "had no hope of becoming a professional man." But the Federal Writers' Project in Chicago prompted him to "develop vocational ambitions" (Davis 12). The South Side Community Arts Center (SSAC) was the first major teaching and exhibition space for Black artists in Chicago, and one that would launch a range of important Black artists, including poet Gwendolyn Brooks and *Life* magazine's first Black photojournalist Gordon Parks. Nat King Cole performed at the SSAC, and Langston Hughes and Paul Robeson gave public lectures there (South Side). The center had to raise money to buy the building, but the WPA paid the salaries of administrators and teachers, funding the daily routine that kept the project going. Maryemma Graham goes so far as to argue that "the largest single impact on black writing before the civil rights movement was really the WPA, not the Harlem Renaissance" (qtd. Nance).

The experience of visual artists points to a similar conclusion. As a young man, Jacob Lawrence took art workshops sponsored by the WPA in Harlem. Sculptor Augusta Savage then encouraged him to get on the federal payroll. It was thanks to the WPA that he was able to focus full-time on his art, painting, learning from other artists, and reading art history and philosophy. Lawrence recalled receiving "the fabulous salary of $23.86 a week" (Nicholas 260). For Eldzier Cortor, similarly, it was the WPA that ensured he "could make money as an artist" (Cortor et al.). Before the WPA, printmaker Dox Thrash had worked as "a porter, elevator operator, buck-and-wing dancer, steward and soldier," but by 1940, *The Crisis* reported, he had become "one of America's leading Negro artists" (Wood 379). Romare Bearden later wrote:

> Paradoxically, it was during the Great Depression of the 1930s that significant numbers of African-American artists were able to work at their art full time for the first time, through the government work-relief art projects .... not only did this legitimate blacks as professional artists, which they had rarely been considered before, it also provided for their artistic training and development. (Bearden and Henderson 227)

Fay Chong, a Chinese-American printmaker and watercolorist in Seattle, counted himself "very fortunate" to have been hired by the WPA. "I know that it helped me a lot, and I think it helped others too ... to get started," he said (Chong and Bestor 4). When asked if he found the government program constraining, he praised the freedom he had to choose his subject matter and did not seem to mind the production schedule: "there was of course a time requirement that we had to produce so many prints or so many paintings in a certain time, you see, but I suppose that is a necessity for a project or for any kind of organization" (6). "I believe it was a requirement of doing one a week or something like that. And we were permitted to work outside, or outdoors, or work in our studio so we were very free at that time, too, which to us was a very good thing" (4). The interviewer pressed him, noting that other artists "felt it was sort of too restricted an atmosphere." Chong responded, unruffled: "that may be, but I didn't feel that way about it" (4).

Edward Arcenio Chavez, a muralist from New Mexico, agreed: "wherever I hear discussions about Federal support of art and the limitations, my experience was that I had more freedom, possibly more freedom, than one does now with coping with the galleries and dealers, and so forth. Complete freedom in what we could do" (Chavez and Trovato 3).

A number of influential women artists, too, got their start thanks to the WPA. Lee Krasner had given up painting to retrain as a teacher, but a paycheck from the FAP Murals allowed her to return to art. "The whole experience taught me scale," she recalled (qtd. Ortega 153). Dorothea Lange became part of a whole new generation of photographers thanks to the Farm Security Administration. "We photographers were somewhat picked at random, we weren't hand-picked. We were educated on the job. The United States Government gave us a magnificent education, every one of us. And I don't know any that's really fallen by the wayside" (Lange and Doud 6).

Award-winning African American poet Margaret Walker struggled to make ends meet after graduating from Northwestern in the depths of the Depression in 1934. She found work with the WPA on the Illinois State Guide, but her writing was so good that she was eventually accorded the freedom to write anything she wanted (Strand 216). Friendly—and later not so friendly—with Richard Wright, Walker was grateful to the project for allowing her to meet other writers, such as Arna Bontemps and Sterling Brown, and to participate in the South Side Writers Group. Her work was also shaped, Jerrold Hirsch explains, by the "Living Lore" project in Chicago, where WPA writers had the "opportunity to explore the relationship between the folklore collection and creative writing processes, functioning much like modern university-based writers' programs" (qtd. Strand 221). Walker incorporated folk stories she heard in interviews with Black people born in the South who had migrated to Chicago (Strand 222). She then spent the rest of her life extolling the WPA for having given many Black artists the infrastructure they needed to work.

To be sure, substantial inequalities also shaped the experience of artists on the federal rolls. Mary Ann Calo's extensive research reveals the small proportion of Black artists employed, the frequent refusals to recognize Black artistic traditions and expertise, and the segregated art galleries across the South, where an urban or regional art center would be established for white artists, followed by a "negro extension." Artists who were not US citizens were thrown off the WPA payroll in 1937, including Arshile Gorky, Mark Rothko, and Willem de Kooning. Censorship shut down a production of *The Cradle Will Rock*, and the Mayor of Chicago blocked play after play (Osborne 20–3). African American sculptor Selma Burke won a commission to create a plaque to FDR but was micromanaged at every step (Calo 106). Editors of the state guides repressed references to historical events, such as the Sacco and Vanzetti trial in Massachusetts, and Sterling Brown, working in Washington, counteracted many false and stereotypical accounts of Black people (Sharps 299). By the late 1930s, the arts programs were so relentlessly accused of being wasteful and spreading communism that the government cut pay and laid off workers. Artists responded with strikes and sit-ins, provoking further attacks from the right. The programs shrank considerably in 1941 and were shut down altogether in 1943.

Despite the many failures of the WPA, I want to draw some lessons from the arts projects about the affirmative affordances of routines as forms. The first and most important is the temporal infrastructure the WPA established for creative work. A wide range of people suddenly got regular blocks of time, preserved from interruptions like a day job or childcare, and freed from worries about making ends meet, to focus on the making of art.

Second, the daily WPA routines allowed artists, in a culture that took labor seriously, to understand art as a dignified and meaningful pursuit. For a number of artists, this meant casting themselves as professionals, rather than amateurs or hobbyists. As painter Herman Cherry put it, "We couldn't believe that you got paid *steady*, twenty-three dollars a week, just to paint. It was the craziest thing we ever heard of" (qtd. Opdycke 90).

The third affordance of laboring routines is perhaps the most surprising. Although we usually associate industrial labor routines with soul-destroying, unthinking obedience to the clock and the machine, WPA work routines enabled the most radical experimentation in art. Even a brief glance at the list of artists on the WPA payroll makes this clear—not only Ellison, Hurston, de Kooning, Lawrence, Rothko, Smith, and Wright, but also Diego Rivera, Louise Nevelson, and Ad Reinhardt, among many others. In some cases, innovations followed from collaborations afforded by shared spaces on the clock. Thrash worked with other printmakers to invent the carborundum printing process, which produces a sandy, chiaroscuro texture still in use by printmakers today. Other artists found inspiration from each other as they waited together on the WPA lines. "All the artists used to go down to project headquarters … to sign in," Lawrence recalled: "We'd meet each other and talk and talk and talk" (qtd. Hills 38).

The fourth and final affordance of the work routine was that it invited artists to understand themselves as workers who were waging a collective struggle alongside carpenters and garment workers for decent pay and conditions. This led to new solidarities. National Maritime Union workers joined artists on the picket line, and artists supported sales-workers on strike in Brooklyn (Lampert 163).

My larger goal here is to revalue routine for both politics and art. As critics and artists have fought vigorously against routine as artificial and enforced, while celebrating rhythms, imagined as synchronized with nature, our bodies, and expressive desires, I have made the case that we are missing the ways that formally speaking, they are the same—organizations of experience into repetitive patterns. And so, we have also missed the crucial importance of routines to the flourishing of collective life.

Because human bodies need nutrients and sleep over and over again, all societies have incorporated repetitive forms into the patterns of basic survival (Levine). But the specific shapes that these routines take vary widely. Food, rest, and labor can be distributed in egalitarian ways, as they are in the cyclical allocations of modern Basque communities (Graeber and Wengrow 295–96); or distributions can be unjust, like the 35 percent of female-headed households that are food insecure in the United States (Rabbitt 20). I opt for the term routine over rhythm, here, to put a deliberate emphasis on the constructedness of these repetitive patterns. All social forms are designed and made, which means that to call some of these rhythms is to mystify—and to entrench—them. To call them routines is to recognize their artfulness and their capacity to be *remade*, not rejecting them out of hand, but putting our energy into building routines for social transformation and even revolution (Pedwell 3–27).

In revaluing routine, I join a tradition of feminist thinkers, including bell hooks, Susan Fraiman, and Marquis Bey, who have argued that critics have made too much of avant-garde ruptures at the expense of the routine labor of ordinary survival, which is typically feminized—and taken for granted. When we talk about radical politics, for example, we often celebrate the public faces of resistance, mostly male: "people who give speeches, negotiate with bosses and politicians, get published, get elected, and otherwise become visible as actors in ways that align with dominant hierarchies" (Spade 135). In this context, it has been easy to forget those, mostly women, who have made resistance possible in the background, by "typing, filing, phoning, feeding, healing, supporting, loving, and occasionally even participating on the front lines" (Sargent ix). Downplaying this mundane work behind the scenes is not only unjust; it is also a strategic mistake. As Dean Spade argues, when communities on the front lines provide regular food and medical care, they draw new participants to political action and help to sustain participation over time. FBI Director J. Edgar Hoover famously called the Black Panthers' free breakfasts for children, transportation, and health clinics "potentially the greatest threat to efforts by authorities to neutralize the BPP and destroy what it stands for" (qtd. Spade 136).

It is particularly important to reclaim routines in this moment of "flexible capitalism," when nearly 90 percent of people in the United States will at some point work nonstandard hours (Suleiman et al. 1). Contingent workers, from hotel housekeepers to adjunct faculty, get little advance notice about changes in scheduling. Not being able to count on regular wages, and having sleep and childcare constantly disrupted, workers "run from part-time job to part-time job and from unpredictable shift to unpredictable shift. They are economically insecure ... but also frazzled from trying to coordinate a chaotic assemblage of irregular schedules" (Snyder 7). At the other end of the wage scale, high earners are also perpetually on call now, too, working long hours in virtual "flexible workplaces" that never allow them to be entirely free from the demands of work. Both salaried and hourly women regularly express the wish that they had more time to spend with their children (Lambert). Unpredictable shifts also disrupt bodily rhythms, including regular sleep, and impede collective rhythms, such as mealtime gatherings.

It turns out that routine itself may never have been the greatest cause of suffering for workers. A recent study of catering workers concluded with a "surprising result":

> When asked about the nature of their jobs and how they felt about them, all workers said that their jobs were repetitive but they were not bored .... One catering assistant said "I don't feel uncomfortable with the routine. You get yourself all set in a different way into a routine." .... Routine seems to bring some measure of order to their work and helps them find security in predictability. One dishwasher said that he liked the routine because it was predictable. "It helps you to do things correctly ... because you always know what to expect ... you know what people expect of you, you know what to do." (Tsai 220–21)

A formalist analysis allows us to disaggregate the different temporal forms that organize labor: we might first think of the *repetitive sequence* of tasks, but work is also shaped by the *rate* or speed of each task, and the *duration* of work without rest. Workers often respond bitterly when management increases the speed of their tasks or the length of the working day, but express enthusiasm for predictable routines, which suggests that fast pace and sustained duration are more painful than routine (Form 4, 12).

Artists are unlikely representatives for the importance of routines. But I have brought art together with routinized work here to resist "the cult of creativity"—the postwar fascination with special individuals, rebellious talents set apart from unthinking conformists (Franklin). This ideology around art-making, as I have suggested, diminishes women's work and the vast mass of other laborers engaged in practices of maintenance. If routine forms are necessary to collective life, we would do better to understand the importance of predictability and to design and fight for better routines. But the humanities today usually do the opposite: we overwhelmingly favor challenging, disruptive, innovative art, and so we miss the crucial political affordances of repetitive rhythms, which include sustaining bodies and minds for the long haul, training us for successful resistance, and drawing people together in regular meetings and mass actions (Levine).

Happily, the story of the WPA allows us to have our revolutionary art and politics as well as our humdrum routines. It shows that all workers, even the most bohemian and experimental outliers, benefit from the predictability of regular food and time to focus. From this perspective, the most monotonous routines afford us the freedom to daydream, plan for revolutionary change, and, yes, make art.

Works Cited

Abate, Jason. *Eager and Hungry for Music: The WPA Music Project in New Orleans, 1935–1943*. University of New Orleans Theses and Dissertations, 2006.

Albro, Maxine, and Parker Hall. Oral history interview with Maxine Albro and Parker Hall. Smithsonian Archives of American Art. July 27, 1964. www.aaa.si.edu/download_pdf_transcript/ajax?record_id=edanmdm-AAADCD_oh_213349.

Asure, Robert, and Harlan Phillips. Oral history interview with Robert Asure. Smithsonian Archives of American Art. October 7, 1965. www.aaa.si.edu/download_pdf_transcript/ajax?record_id=edanmdm-AAADCD_oh_213238.

Attridge, Derek. *The Singularity of Literature*. Routledge, 2004.
Baraka, Imamu Amiri. *Black Magic: Sabotage, Target Study, Black Art: Collected Poetry, 1961–1967*. Bobbs-Merrill, 1969.
Barletta, Vincent. *Rhythm: Form and Dispossession*. U of Chicago P, 2020.
Bearden, Romare, and Harry Brinton Henderson. *A History of African-American Artists: From 1792 to the Present*. Pantheon, 1993.
Bey, Marquis. *Black Trans Feminism*. Duke UP, 2022.
Biddle, George, and Harlan B. Phillips. Oral history interview with George Biddle. Smithsonian Archives of American Art. 1963. www.aaa.si.edu/download_pdf_transcript/ajax?record_id=edanmdm-AAADCD_oh_213269.
Bindas, Kenneth J. *All This Music Belongs to the Nation: The WPA's Federal Music Project and American Society*. U of Tennessee P, 1995.
Blakey, George T. *Creating a Hoosier Self-Portrait: The Federal Writers' Project in Indiana, 1935–1942*. Indiana UP, 2005.
Boyd, Valerie. *Wrapped in Rainbows: The Life of Zora Neale Hurston*. Simon and Schuster, 2003.
Brown, Frederick. *Flaubert: A Biography*. Harvard UP, 2006.
Bunce, Louis, and Dorothy Bestor. Oral history interview with Louis Bunce. Smithsonian Archives of American Art. October 29, 1965. www.aaa.si.edu/download_pdf_transcript/ajax?record_id=edanmdm-AAADCD_oh_21333.
Calo, Mary Ann. *African American Artists and the New Deal Art Projects*. Penn State UP, 2023.
Chavez, Edward, and Joseph Trovato. Oral history interview with Edward Chavez. Smithsonian Archives of American Art. November 5, 1964. www.aaa.si.edu/download_pdf_transcript/ajax?record_id=edanmdm-AAADCD_oh_213362.
Chong, Fay, and Dorothy Bestor. Oral history interview with Fay Chong. Smithsonian Archives of American Art. February 14–20, 1965. www.aaa.si.edu/download_pdf_transcript/ajax?record_id=edanmdm-AAADCD_oh_213741.
Clark, Shannan. "When Modernism Was Still Radical: The Design Laboratory and the Cultural Politics of Depression-Era America." *American Studies*, vol. 50, nos. 3/4, fall/winter 2009, pp. 35–61.
Cortor, Michael, Sarah Kelly Oehler, and Mark Pascale. "Eldzier Cortor through the Eyes of His Son." Art Institute. 2020. www.artic.edu/articles/879/eldzier-cortor-through-the-eyes-of-his-son.
Currey, Mason. *Daily Rituals: How Artists Work*. Doubleday, 2013.
Currey, Mason. *Daily Rituals: Women at Work*. Knopf Doubleday, 2019.
Davis, Thadious. "Becoming: Richard Wright and the WPA." *The Black Scholar*, vol. 39, nos. 1/2, spring/summer 2009, pp. 11–16.
Dewey, John. *Art as Experience*. Perigee Books, 1980 (1934).

Du Von, Jay, and Harry Hewes. Oral history interview with Harry Hewes and Jay Du Von. Smithsonian Archives of American Art. October 1964. www.aaa.si.edu/download_pdf_transcript/ajax?record_id=edanmdm-AAADCD_oh_213568.

Ellison, Ralph. *Going to the Territory*. Knopf Doubleday, 2011 (1986).

Emerson, Ralph Waldo. *Essays and Lectures: Nature, Addresses and Lectures*. Library of America, 1983.

Engels, Friedrich. *The Condition of the Working Class in England in 1844*. Translated by Florence Kelley Wischnewetsky. Swann Sonnenschein, 1892.

Flaubert, Gustave. *Œuvres complètes de Gustave Flaubert: Correspondance*. 4th series. Louis Conard, 1905.

Form, William H. "Auto Workers and Their Machines: A Study of Work, Factory, and Job Satisfaction in Four Countries." *Social Forces*, vol. 52, no. 1, September 1, 1973, pp. 1–15.

Fraiman, Susan. *Extreme Domesticity*. Columbia UP, 2017.

Franklin, Samuel W. *The Cult of Creativity: A Surprisingly Recent History*. U of Chicago P, 2023.

Fraser, Max. Interview with Molly Morrow, "A More Perfect Union: American Political Art of the 1930s." Metropolitan Museum of Art. September 22, 2023. www.metmuseum.org/perspectives/fraser-political-art-of-great-depression-interview.

Graeber, David, and David Wingrow. *The Dawn of Everything: A New History of Humanity*. Farrar, Straus, and Giroux, 2012.

Hapke, Laura. *Labor's Canvas: American Working-Class History and the WPA Art of the 1930s*. Cambridge Scholars, 2008.

Hills, Patricia. *Painting Harlem Modern: The Art of Jacob Lawrence*. U of California P, 2009.

Hirsch, Edward. "Derek Walcott: The Art of Poetry." *The Paris Review*, vol. 101, winter 1986, pp. 196–230.

hooks, bell. *Feminist Theory: From Margin to Center*. South End, 1984.

Jackson, Lawrence P. *The Indignant Generation*. Princeton UP, 2021.

James, William. *The Principles of Psychology*, vol. 1. Henry Holt, 1905.

Jeansonne, Glen. *A Time of Paradox*. Rowman and Littlefield, 2007.

Kainen, Jacob. "The Graphic Arts Division of the Federal Art Project." *The New Deal Art Projects: An Anthology of Memoirs*. Edited by Francis C. O'Connor. Smithsonian Institution P, 1972, pp. 155–76.

Kennedy, Stetson. "Florida Folklife and the WPA." 1990. www.floridamemory.com/learn/classroom/learning-units/zora-neale-hurston/documents/stetsonkennedy/.

Kimmelman, Michael. "Wake Up. Wash Face. Do Routine. Now Paint." *New York Times*, May 8, 2005, section 2, p. 1.

Kuehn, Manfred. 2001. *Kant: A Biography*. Cambridge UP, 2001.

Lambert, Susan J. "When Flexibility Hurts." *New York Times*, September 19, 2012, www.nytimes.com/2012/09/20/opinion/low-paid-women-want-predictable-hours-and-steady-pay.html.

Lampert, Nicholas. *A People's Art History of the United States.* New P, 2013.

Lange, Dorothea, and Richard K. Doud. Oral History Interview with Dorothea Lange. Smithsonian Archives of American Art. May 22, 1964. www.aaa.si.edu/download_pdf_transcript/ajax?record_id=edanmdm-AAADCD_oh_213615.

Lefebvre, Henri. *Rhythmanalysis: Space, Time, and Everyday Life.* Translated by Stuart Elden and Gerald Moore. Continuum, 2004.

Levine, Caroline. *The Activist Humanist: Form and Method in the Climate Crisis.* Princeton UP, 2023.

Lynch, Tosca A. C. "Rhythmics." *A Companion to Ancient Greek and Roman Music.* Edited by Eleonora Rocconi and Tosca A. C. Lynch. Wiley, 2020, pp. 275–95.

Maclagan, David. *Outsider Art: From the Margins to the Marketplace.* Reaktion, 2009.

Meyers, Mary Anne. *Albert Barnes and the Science of Philanthropy: Art, Education, and African-American Culture.* Routledge, 2017.

Miller, Arthur I. *Einstein, Picasso: Space, Time, and the Beauty That Causes Havoc.* Basic Books, 2008.

Moss, T. G., et al. "Is Daily Routine Important for Sleep?" *Chronobiology International,* vol. 32, no. 1, 2014, pp. 92–102.

Musher, Sharon Ann. *Democratic Art: The New Deal's Influence on American Culture.* U of Chicago P, 2013.

Nance, Kevin. "Roosevelt's Writers." *Poets & Writers,* July/August 2010. www.pw.org/content/roosevelt039s_writers.

Nicholas, Xavier. Interview with Jacob Lawrence. *Callaloo,* vol. 36, no. 2, spring 2013, pp. 260–67.

Opdycke, Sandra. *The WPA: Creating Jobs and Hope in the Great Depression.* Routledge, 2016.

Ortega, Jesse Laird. "Chronology." *Women of Abstract Expressionism.* Edited by Joan Marter. Yale UP, 2016, pp. 153–61.

Osborne, Elizabeth A. *Staging the People: Community and Identity in the Federal Theatre Project.* Palgrave Macmillan, 2011.

Palma, Rubén. "Helena Minginowicz—The Strength of Fragility." *Overstandard,* January 7, 2025. https://overstandard.dk/helena-minginowicz-the-strength-of-fragility/.

Pedwell, Carolyn. *Revolutionary Routines.* McGill-Queen's UP, 2021.

Pierucci, Antone R. E. *The Story of Leonardo Da Vinci.* Atlantic Publishing, 2017.

Pillen, Cory. *WPA Posters in an Aesthetic, Social, and Political Context.* Routledge, 2020.

Rabbitt, Matthew P., et al. "Household Food Security in the United States in 2023." USDA Economic Research Report Number 337. September 2023. https://ers.usda.gov/sites/default/files/_laserfiche/publications/109896/ERR-337.pdf?v=88718.

Sargent, Lydia. "New Left Women and Men: The Honeymoon is Over." *Women and Revolution.* Edited by Lydia Sargent. Black Rose Books, 1981, pp. ix–xxx.

Schott, Webster. "Autumn for 'The Age of Anxiety.'" *Life,* January 30, 1970, pp. 52–5.

Shahn, Ben, and Harlan Phillips. Oral history interview with Ben Shahn. Smithsonian Archives of American Art. October 3, 1965. www.aaa.si.edu/download_pdf_transcript/ajax?record_id=edanmdm-AAADCD_oh_213824.

Sharps, Ronald LaMarr. *Black Folklorists in Pursuit of Equality*. Lexington Books, 2023.

Shelley, Percy Bysshe. "A Defence of Poetry." *The Prose Works*, Vol. 3. Edited by Harry Buxton Forman. Reeves and Turner, 1880 (1821).

Shklovksy, Viktor. "Art as Technique." *Russian Formalist Criticism: Four Essays*. Translated by Lee T. Lemon and Marion J. Reis. U of Nebraska P, 1965, pp. 16–30.

Simpson, Kathryn. "Friends and Lovers." *The Oxford Handbook of Virginia Woolf*. Edited by Anne E. Fernald. Oxford UP, 2010, pp. 27–43.

Snyder, Benjamin H. *The Disrupted Workplace*. Oxford UP, 2016.

Solman, Joseph. "The Easel Division of the WPA Federal Art Project." *The New Deal Art Projects: An Anthology of Memoirs*. Edited by Francis C. O'Connor. Smithsonian Institution Press, 1972, pp. 115–30.

Solomon, Deborah. *Jackson Pollock: A Biography*. Cooper Square Press, 1987.

Soltan, Jerzey. "Working with Le Corbusier." *Le Corbusier Archive*, vol. 17. Edited by H. Allen Brooks. Garland, 1983, pp. ix–xiv.

South Side Community Arts Center. "About Us." https://sscartcenter.org/about-us/.

Spade, Dean. "Solidarity not Charity: Mutual Aid for Mobilization and Survival." *Social Text*, vol. 38, no. 1, March 2020, pp. 131–51.

Strand, Eric. "Margaret Walker and the WPA: Black Feminism, Progressive Government, and the Program Era." *ELH*, vol. 91, no. 1, spring 2024, pp. 207–38.

Suleiman, Adekami O., et al. "Worker Perspectives on the Impact of Non-Standard Workdays on Worker and Family Well-Being." *BMC Public Health*, vol. 21, no. 2230, December 2021, pp. 1–12.

Swain, Martha H. "Women's Work Relief in the Great Depression." 2004. mshistorynow.mdah.ms.gov/issue/womens-work-relief-in-the-great-depression.

Sylvester, David. *Interviews with American Artists*. Yale UP, 2001.

Tate, Claudia. "Maya Angelou: An Interview." *Maya Angelou's I Know why the Caged Bird Sings: A Casebook*. Edited by Joanne M. Braxton. Oxford UP, 1999, pp. 149–58.

Taylor, Nick. *American-Made, The Enduring Legacy of the WPA*. Random House, 2009.

Thompson, Jerry L. *The Last Years of Walker Evans*. Thames and Hudson, 1997.

Trelawny, Edward. *Records of Shelley, Byron, and the Author*, vol. 1. Benjamin Bloom, 1968 (1868).

Tsai, Chin-Ju. "Boredom at Work and Job Monotony: An Exploratory Case within the Catering Sector." *Human Resource Development Quarterly*, vol. 27, no. 2, summer 2016, pp. 207–36.

Wood, Jerry. "A New Print Process." *The Crisis*, vol. 47, no. 11, December 1940, p. 379.

Part Three

Aesthetic Manifestations

8

The Measures of Fugitive Time

Keith D. Leonard

In *Suddenly We*, the title phrase of her most recent poetry book, poet and scholar Evie Shockley has captured the compelling ideal of aesthetic experience through which she and several of her contemporaries have attempted the impossible. Her title's proposition is of a serendipitous communion, an almost ecstatic realization of solidarity across lines of difference brought to one's momentary attention by art. Although this claim sounds like traditional aesthetic identification, it is distinct in its resistance to diminishing a social "other" into a static, passive external enactment of the self, thereby resisting the terms of enforced sameness by which Black writers like Shockley have historically been praised for "transcending" race by disclaiming aspect of themselves. Shockley's title posits instead an experience of equalizing gestalt that preserves what Edouard Glissant would call "irreducible singularity," the unconsumable difference, the "opacity," that construes identification as recognizing, accepting, and respecting what we cannot know about others at least as much as what we presume to know (190). Moreover, this equalizing encounter of difference is a function of its suddenness, an experience of neither the atemporal nor the teleological conceptions of modern subjectivity associated with the lyric (Jackson, *Dickinson's Misery*), but rather a brief instance of prolepsis, the anticipation of realized, if partial, union made possible in innovative form. Against a long heritage of African diasporic poets insisting, often in metered, conventional poetics, that Europeans and others identify with the basic humanity denied to them, Shockley and her peers aspire to generate aesthetic experiences of human unity framed from a Black perspective that refuse to prioritize the predominant racial ideologies that reduce identification to sameness, which amounts to a cultural erasure, what has come to be called a "social death." They seek instead to render fuller articulations of Black artistic self-comprehension as viable sources of human connection.

What Shockley's brilliant title illuminated for me, in other words, is that innovative African American poets like her persuasively complicate the conventional, universalized temporality of imagined connection through aesthetic experience by manipulating structures of poetic time to produce senses of this proleptic gestalt. Shockley and fellow innovators such as M. Nourbese Phillip, Dionne Brand, Robin Coste Lewis, and Nathaniel Mackey adapt poetic rhythm, sometimes but not always in relation to music, in the name of the possible, of the not-yet-realized, through poetic structures that invite but then defer or disrupt musical, logical, and temporal closures. The techniques they deploy, including rhythmic phrasing rather than meter, complex visual layouts aligned with that phrasing, and densely layered semantic play, render any alignments of sound patterns with claims to selfhood productive but temporary, calling attention to their constructedness to imply multiple other possible alignments. I call these and other techniques "measures of fugitive time" to indicate both the poets' musical phrases structuring their verse in various time signatures, so to speak, and their implicit sense that rhythmic structures are at their best when they admit being inadequate or partial gauges of temporal experiences, a purposeful inadequacy that proposes to structure a collective and momentary access to what has come to be theorized as Black "fugitivity." The "fugitive" consists of modes of nonobjectified Black being that resist by exceeding the Manichean logic of a Western liberalism predicted in foundational ways on its opposition to Blackness. This "fugitivity" encompasses distinctive conceptions and experiences of time, both in resistance to notions of linear progress in which African and Black diasporic cultures lag behind European ones in evolution toward modernity, and in utopian projections into better possible futures, whether those futures are immediately pursuable or not. Deploying time in the name of this fugitivity through well-crafted metrical disruptions produces ongoing anticipation rather than claims to final fulfillment, challenging the totalized or enclosed modes of subjectivity and identification produced by the logics of racial opposition in which aesthetic sameness presents itself as the most substantive mode of mutual recognition within a modernity defined by the exclusion or diminishment of blackness. These poets rightly and persuasively imagine otherwise. Expanding upon the implications of these measures only implicit in Shockley's title, I seek to illuminate how these poets posit the fullness of Black being as a(n im)proper if partial measure of human being itself, and therefore as a more than adequate dissent from modernity's erasures.

I trace these measures of fugitive time, then, because this sense of sudden, momentary, and proleptic communion constitutes one of the most effective recent answers to a key question about art, race, and identification that has long troubled African American literati. In her essay "On Seeing and Reading the 'Nothing': Poetry and Blackness Visualized," Shockley rightly asks, "what can we articulate about blackness in a colonial language, particularly English, that is not encrusted with Europe's ancient equation of blackness with evil and fear, now wedded to the modern era's demonizing discourses of race" (500)? Jared Sexton, Frank Wilderson, Saidiya Hartman, Fred Moten, Christina Sharpe, and Matthew Omelsky, among other recent purveyors of the Black radical tradition's "ontological thinking," have identified well the long-standing intellectual foundations of this question. They have clarified through their studies of Western philosophy how the coherence of the liberal humanist subject has been predicated on its opposition to Blackness objectified in the abstract, a conceptual coherence secured in concrete practice through the enslavement, segregation, and cultural erasure of people racialized as Black. In these terms, the "social death" that Orlando Patterson had originally and narrowly argued was the defining social status of the enslaved in all known practices of slavery attaches instead to Blackness not only as a social but as a more fundamental ontological condition. This conclusion derives in part from how often Enlightenment thinkers explicitly excluded Africans from their theories of "universal humanity" and how consistently social inequality in Europe and the United States is racialized accordingly. Soon after, as Virginia Jackson has argued, traditional lyric identification became conflated with the modern subjectivity constituted in these ways against Blackness, effectively reducing lyric identification to the mirroring in all readers of the bourgeois, heterosexual, European-American Christian male perspective to the exclusion of other possible subjects identifying in other possible ways. Sharpe, Hartman, Kevin Quashie, and Anthony Reed turn to poets such as Phillips, Brand, and Lewis to confirm how fully and directly innovative literary practices by African diasporic poets chronicle, reveal, and reckon with these deadening effects. But what has not been acknowledged nearly enough, and what Shockley emphasizes, is that these poets also seek to make sensible the allegedly "impossible" Black social life within and against the terms of human subjectivity so "encrusted." Their work tends to the "presence" necessarily awaiting somewhere "under" that crust for some kind of articulation.

Where Shockley explores how visual poetics constitutes a key response to this problem, I propose that nontraditional deployments of rhythm may constitute an even more powerful feature of that articulation. The chief innovation of this cadre of poets, I thus contend, is their sense that innovative poetic form in general can simultaneously register the depredations of racial discourse, as Black ontological thinking rightly critiques, *and* instantiate a sense of an African diasporic consciousness or self-comprehension operative within and defiant of that encrusted discourse. Their senses of rhythm in particular propose an alternative to the binary oppositions of social death, a synthesis, or at least a simultaneity, that promises greater appreciation of Black self-comprehension, no matter how encrusted, through its capacity to recognize itself and be recognized by others. This simultaneity is perhaps best understood as sensing (not only seeing) one's existence from at least two places at once, an awareness that rhythm facilitates by oscillating between anticipated closure and aspirational disruption. That oscillation helps to shift comprehension from visual metaphors tied too closely to the rational policing of difference and such static oppositions as presence and absence to those of sound, movement, and even touch that manifest meaning and being in more fully blurred boundaries. Even with her emphasis on the visual, Shockley captures this possible simultaneity well: "How can black subjectivity—not the *object seen* as black, but *what black-identified people see through the lens of their blackness*—be expressed in a language developed over the past few centuries precisely to facilitate, legitimate, justify, and downplay the commodification and devaluing of the bodies who might speak 'in the first person' about this perspective" ("On Seeing" 500, original emphasis)? In answer, poets like Shockley and thinkers like Sharpe, Quashie, and Moten amplify the fugitivity of this Black self-awareness, its possibly validating excess of objectification available through its very otherness, what Moten described as "a fugitive movement in and out of the frame, bar, or whatever externally imposed social logic—a movement of escape, the stealth of the stolen that can be said, since it inheres in every closed circle, to break every enclosure." Elaborating on Heidegger's distinction between object and thing, Moten describes the "stolen" as "the ones who inhabit the nearness and distance between *Dasein* and things (which is off to the side of what lies between subjects and objects)" (186). Interested in this space off to the side too, Shockley and her peers imply that this "stealth of the stolen" resides in this distinctive and differing self-comprehension that, because it has been conceived by Western thought as the external boundary of the sensible, must derange English grammar, poetic rhythm, and associated senses of temporality to make the most sense of itself.

But, contrary to the most pessimistic senses of the totality of social death, it can indeed make sense of itself, as Shockley wisely observes. "Given poetry's unique position as an art made with words that is also fundamentally and self-consciously concerned with imagery … and spatiality," Shockley declares, "Poetry thus recommends itself *formally* for the aesthetic project of rendering black subjectivity as image-text—and, perhaps, epistemologically as well, to the extent that poetry is almost as impossible to pin down generically as blackness is conceptually or experientially" ("On Seeing" 500, original emphasis). If we add temporality to Shockley's conjunction of "image-text," we can comprehend even more fully how innovative poetry, as an uncategorizable compendium of verbal–visual strategies, can indeed match well with the uncontainable contradictions, alibis, violence, affirmations, confusions, and productive misalignments summoned by Blackness as color, racial construct, and cultural name. Poetic rhythm can make sensible established patterns and the counterpatterns available in those patterns and can deploy closure alongside resistance to closure, using the specific anticipated rhythms of Black music and speech in particular as basis and departure. Rhythm in these practices is of course associated with the patterns of syllables in language, but metrical pattern and counterpattern together can also claim to align with the allegedly natural rhythms heard or felt as breath and heartbeat and even patterns of thought. The compelling claim is that disjunctions and disruptions of innovative poetic time align with how those life patterns shift, pause, accelerate, and defibrillate in relation to one another, making sensible that which is otherwise conceptually denied. Such poetics call on all of the human senses to attend to what might not seem to be there, the Black being, the thingliness of Blackness, off to the side of subjects and objects. Understood with such techniques and aims in mind, Shockley's title phrase does indeed usefully frame a poetic aspiration toward a provisional community (or at least communion) available in fugitive aesthetic assertions of Black imaginative being.

The burden of this chapter is thus to exemplify through Shockley and fellow innovator Nathaniel Mackey some of the practices by which contemporary African American poets pursue this breaking of racialized ontological enclosures—to borrow perhaps an overstatement from Moten—by acknowledging both what those enclosures forcefully claim to contain and erase *and* what of being exceeds them. Although Shockley leans more toward social critique and protest than Mackey, who asserts a utopian spirituality he may not believe in, both evoke how multisensory self-awareness exceeds social death by capturing the purposefully dissonant movements of that provisional self-comprehension in unconventional poetic rhythms associated with improvisational Black music and produced in part by puns, allusion, anagrams, and visual cues. To make this case, I analyze Shockley's poem "post-white" from her volume *the new black* (2012) to identify how her rhythmic parody of the anthem "America" punctures US patriotic culture to imply what an adequately inclusive American identity might sound like. I then turn to the first sections of Mackey's lifelong "Mu" series from his first two books, *Eroding Witness* (1985) and *School of Udhra* (1993), to trace how Mackey more explicitly imagines a utopic communion through gnostic possibility, the glimpses of something that one might call a shared human spirit potentially available through mystic rituals which are replicated in verse forms through disjunctive rhythmic principles inspired by free jazz. In each case, the poet posits rhythmic disruption or resistance to closure as an access point to different ways to move through time toward a more inclusive, open-ended spirit of togetherness, whether it is turning "America" into a blues song that writes the afterlives of slavery into patriotic becoming, or portraying multiply identified communities—Mackey's oft-cited amorphous "we"—engaging in rituals of meaning from worldwide religious traditions that Mackey unites through musical techniques associated with Blackness to structure the search for what "we" might mean.

I contend that these interventions engage in "fugitive time" by seeking to exemplify in their structures the real and aspirational temporalities allegedly generated within the erasures produced by social death and its allegedly alienating Western "clock time." I identify how Shockley's and Mackey's practices structure a possible experience of the "*duree*" that, as Omelsky derives it from Frantz Fanon, "corresponds to the temporal and affective movement from alienation to disalienation, from nonbeing to being. More precisely, it's the *anticipation* of disalienation" (23, original emphasis), "the unanticipatable world of relief [from oppressive pain] that, impossibly, must be dreamed in advance" (25). One key difference I emphasize is that these poets do not seem to accept Black nonbeing or its inevitable future as fully as Omelsky, Sexton, Wilderson, and others do. Instead, fashioning that being and its articulated self-comprehension as a proleptic, unifying, fleeting temporality, Shockley and Mackey exemplify a Black tradition's "insistence on existing" (Sharpe 11) that has always been somewhere at some time, however encrusted by the conceptual necessity of its death. Joining Sharpe, Hartman, Quashie, and Zakkiyah Iman Johnson especially in exploring spaces of possibility some "black ontological thought" has foreclosed, I conclude that these two poets achieve a sense of this insistence by effectively structuring in verse form the sudden, provisional, momentary realizations that "we" are sometimes and genuinely the "we" we make ourselves out to be.

The case for this dissident communion starts with refracting the operating notions of recognition through sameness central to conventional conceptions of aesthetic identification, including in a great deal of African American formalist poetic production since the era of enslavement. Such operating terms mark theories of poetic rhythm and characterizations of Black vernacular cultures with scripts of sameness that erase singularity and that often categorize senses of movement in time in terms of absolute racial difference and hierarchy. One can see how both the affirming claims and the debasing stereotypes about Black people naturally having rhythm align with notions of the "throng" asserted in the influential theories of poetic rhythm associated with Francis Gummere. In her introduction to *"This is How We Flow": Rhythm in Black Cultures*, for example, editor Angela M. S. Nelson declares, "Rhythm is *the* textual element that allows us to see and understand best African creative expression" (1, original emphasis) because, "Although rhythm is an important concept in all forms of music, it holds a unique position in traditional African thought" (2). Nelson's characterization sounds a lot like what Gummere thought of as "the dancing feet of the primitive 'throng,' residual in modern meters," a residue that for Gummere remains inferior to modernity because, as Jackson puts it, "poetry's appeal is not just primitive and somatic, but a 'consent of sympathy' that civilizes, that has the agency to transform primitive, affective somatic response into 'the noblest thing in all human life'" ("The Cadence of Consent" 94). Rhythm evolves into "the noblest thing" dubiously through an allegedly precolonial authenticity that endures in the postcolony on the one hand that, on the other, undergirds the logic of liberal humanist progress in which European culture alone embodies that nobility at the end of history. Clearly Nelson and the scholars she includes in her volume want to challenge this teleology by valorizing distinctive African cultural practices that both preceded and endured beyond colonialism, but her terms, like Gummere's, reside in the presumption that Black being constitutes an absolute "other" to liberal humanist subjectivity, which reinforces the dubious and death-inducing opposition between Blackness and modernity. Poets like Mackey and Shockley usefully trouble the grounding logics of this unavoidable opposition.

However, instead of rejecting altogether distinctive ethnic cultural heritages or the potential nobility of the highest modes of human connection, as contemporary ontological theorists like Sexton or Wilderson might suggest, we should recognize with these poets that these concepts can be reconfigured in liberating ways. One can see in their work a comprehension of the distinctive roles of rhythm in Black diasporic cultures presented as multiple and constructed rather than as homogeneous, as well as their sense that rhythm enacts kinds of consent, although they locate that consent less in a civilizing Eurocentric teleology or a differentiating Africanist authenticity than in a sense of multiple and provisionally patterned movements among diverse subjects in scattered moments in time. One result is a more open-ended sense of the unity available in aesthetic experience and a broader provenance for its cross-racial social effects. In one interview, Shockley observes, "I feel especially strongly that one of the strengths of poetry is making connections—or making connections visible and, therefore, one of the ways we can use our art to promote social justice is by writing in ways that help people see that the small pieces of the puzzle we're each focused on all come together in the big picture." She declares in another interview, "In this book [*Suddenly We*] in particular, I was trying to think of how I can do what I do, which is to write as a Black woman, but in a way that invites as many people into this *we* as possible." Her approach aligns with Mackey's suggestion that "Maybe 'union' is a better word than 'unity,' better stressing the act or the incidence of convergence, momentary conjunction or the movement toward conjunction—intermittent, hit and miss." He added, "Maybe the unity of turning, the persistence of turning, is the unity my work is after or at least the unity it wants to make its peace with." I gather from these claims that each poet aspires to a provisional comprehension of a whole made momentarily sensible enough in some of its parts to suggest ongoing grandeur that cannot be named totally either only in authenticity or in a civilized nobility, or any other totalizing discourse for that matter, and thus not subject to the deadening racialist assumptions associated with the colonizing will to total comprehension. And yet the productive residue of each model remains in and through the Black "lens"—or, more aptly, the Black "sensing" of being, they provisionally structure.

In what follows, I attend to the more intricate and capacious conceptions of Blackness tied to the more flexible notions of consent and community available through the more ductile understandings of individual subjectivity by which these poets constitute this Black poetic "sensing." I follow Reed in particular when he describes his chapter on free jazz improvisation as offering "claims that music and poetry make for forms of experience beyond those associated with the sovereign intending subject that dominates Western philosophy" (Reed 144). He adds that Black jazz music developed its expansive critique of that intending subject and its narrowed modes of knowing in a "transitional" postcolonial moment in which "reconfigurations of power/knowledge gave the relationship between embodiment, subjectivity, and non-subjective forms of experience such as ecstasy renewed salience" in an effort to "reconceive … subjective experience itself" (145). He also offers this wise caution: "Rather than parsing the degree to which particular practices are black or European (or affirm blackness or Europeanness), we might uncover truly transgressive aesthetic acts" (163). Similarly liberating himself from the impulse to measure difference too precisely, Quashie clarifies one such "truly transgressive" aesthetic act: "And yet we might suppose that every black text rests on a quiet premise of black humanity—that the text and its aesthetics *assume* being" (2, original emphasis). Assuming Black being complicates the concept of an "intending subject" developed in opposition to such an assumption by attributing intention—or at least motivating source—to an absence made present in any text structuring ("assuming") that absence as source. I trace how Black rhythmic cultures, when adapted to verse form in this framing, can structure this simultaneity of absence and presence, being and nonbeing, what Moten might call "entanglement," to imply relations among "intending subjects" less beholden to the unitary, the bounded, the hierarchical, the binary, even the "intended," available in such "non-subjective" experiences. "Ecstasy" might therefore indicate the best meaning of "fugitivity" in aesthetic experience, a momentary sensing of totality that isn't totalizing because it attends to irreducible singularities, which is to say what cannot be known, making it the possible basis for both intraracial and interracial unities unthought of because not registered in thought.

Shockley's "post-white" is a good place to start for these considerations because it offers a version of the Black ontological critique to gesture toward the fugitivity that such critiques did not originally acknowledge. The prose poem opens "my country tears of thee sparkling on a stiff gray bow tied against cognitive dissonance getting (a)head holding it together keeping it really warm." Instead of singing of America, this poem sings of the "tears" "sparkling" on the bow of a ship that hints at the middle passage of Africans into enslavement while also dripping on the bow tie of respectability politics, the ways in which newly freed and upwardly mobile Africans such as Frederick Douglass, Harriet Jacobs, and W. E. B. Du Bois, say, donned the clothing of respectable decorum as a sartorial aesthetic identification. The quoted passage also alludes to the slum-dwelling "we" of Gwendolyn Brooks's 1945 poem "kitchenette building," characters united in their speculation about whether a dream could survive in their cramped spaces even if they "had time to keep it warm." They "wonder. But not well! not for a minute! / Since Number Five is out of the bathroom now, / We think of lukewarm water, hope to get in it." This allusion contends with the traumas causing the tears by amplifying what Brooks's poem proposes, which is that an anticipated future fulfillment within and against the trauma potentially may be available in or symbolized by the mundane desire for a warm shower.

Indeed, Shockley knows that scholars debate whether Brooks's poem presents humanized or dehumanized figures, making that allusion a way to crystallize the dialectical conundrum at the heart of the ontological critique. In *Scenes of Subjection: Terror, Slavery, and Self-Making in the Nineteenth Century*, her study of newly freed African Americans, for example, Hartman observes, "the question persists as to whether it is possible to unleash freedom from the history of property that secured it, for the security of property that undergirded the abstract equality of rights bearers was achieved, in large measure, through black bondage" (119). If liberal humanist sovereignty is indeed tied so fully to property, and if Blackness made such propertied sovereignty possible by being most fully what it was not, as well as by generating its validating wealth, how could a Black person be a person, let alone a free person? Consequently, the "freedman" inhabited a "burdened individuality," says Hartman, characterized by "being freed from slavery and free of resources, emancipated and subordinated, self-possessed and indebted, equal and inferior, liberated and encumbered, sovereign and dominated, citizen and subject" (117). Simplifying these dialectics, Wilderson is blunter: "the same formative relation of structural violence that maintained slavery remains—upheld explicitly by the police (former slave catchers) and white supremacy generally—hence preserving the equation that Black equals socially dead" (9). Attending to these dynamics, "post-white" acknowledges what Brooks's poem proposes, which is that if a dream, as an expression of such burdened being, cannot withstand the deprivation in which a lukewarm shower instead of a cold one counts as fulfillment, then these figures may be dead indeed, perhaps even to themselves, their possible futures foreclosed for them even in their own minds.

But Shockley's poem ultimately prioritizes the fugitivity of Black being insisting itself in the capacity to wonder itself as that capacity manifests in cultural practices, even ambivalent quotidian ones. Here Shockley's poetic approach aligns with Hartman's description of her own scholarly method in *Wayward Lives, Beautiful Experiments: Intimate Histories of Social Upheaval*: "The aim is to convey the sensory experience of the city and to capture the rich landscape of black social life. To this end, I employ a mode of close narration, a style which places the voice of the narrator and character in inseparable relation, so that the vision, language, and rhythms of the wayward shape and arrange the text" (xiii–xiv). Shockley's poem hints at such arranging rhythms of the unacknowledged less by shared narrative voice than by rich cultural allusion, by revising the next line of "America" with reference to cultural objects in which one can find imbedded instances of the perspectives and sensory experiences Hartman pursues in her close narration. Shockley's poem reads, "sweet land of basketball, barbecue, sweet potato pie and cadillac liberty crowned lady with a torch in her voice." Here objects of Black culture are presented as metonyms for the being of the people whose grilling and baked goods are as much in dialectical belonging in "America," song and country, as baseball and apple pie. On one hand, the consumerism associated with a fine automobile indicates the ambivalence about the extent to which social mobility could be "liberty," an ambivalence that Hartman captures well in *Scenes of Subjection*, quoting Marx: "The transactional agent appeared less as the self-possessed and willful agent than as 'someone who has brought his own hide to market and now has nothing to expect—but a tanning'" (120). Yet Shockley also alludes to the fugitive pride in the resourcefulness necessary to acquire that fine car against all Jim Crow segregationist odds, evoking poems such as Brooks's "Bronzeville Man with a Belt in the Back" or Sterling Brown's "Sporting Beasley" in which characters find validation in being well-dressed, whether conventionally like Du Bois or in dissent like Brown's zoot-suiter.

In this context, Lady Liberty becomes a singer of torch songs whose yearning and pride likewise exceed the market, as her uncommodified unarchived life rhythms testify to futures contravening what "America" claims to contain. Shockley expects (some of) her readers to know that the most famous torch songs were popularized by Black blues women such as Billie Holiday, Sarah Vaughan, and Ella Fitzgerald. "post-white" thus anticipates Sharpe who, after calling the historical "afterlives" of slavery's foundational antiblackness a "wake," seeks to characterize the "rupture" in social death produced by the incontrovertible fact that "we, Black people everywhere and anywhere we are, still produce in, into, and through the wake an insistence on existing: we insist Black being into the wake" (11). In Shockley's reading of Brooks, this is the work that "wonder" does. However much one must acknowledge the appropriation of the blues and jazz at the expense of the tanned hides of those who originated the music anonymously, unpaid, and forgotten, one can nonetheless also imagine a Black Lady Liberty finding in those songs of sadness, bitter humor, and resilience, communal unity and existential validation, a fuller notion of freedom for Black people rooted in communalism, affect, and yearning rather than in static individualism and property, in the insistence of being in song after song.

Shockley amplifies this thematic displacement of liberty from the statue onto insistent Black being through distinctive uses of musical rhythm. The song "America" operates in triple meter—three strong beats per measure—which produces stresses on "sweet land of." This meter resists the iambic rhythm available in the phrase "of liberty," thus pitting two modes of rhythmic closure against one another. Rejecting any closure, Shockley inserts between "of" and "liberty" a series of words and phrases that hint at but also defer alternative rhythms. "Basketball" and "barbecue" each have three syllables with the first one stressed, a parallel amplified by the "b" consonance. Although "cadillac" seemingly repeats the pattern, "sweet potato pie" diverges, as "sweet potato" is four syllables not three, which reminds us that "and cadillac" is four syllables too. Here, the expectation of an alternative closure is subverted by the differing, deferring rhythms of these objects of Black culture inserted into "America." If "America" is a torch song rather than an anthem, a yearning song for a lover who perhaps never was but who perhaps could be, who better to embody it than a Black blues singer whose individuality is so burdened? And what better way to suggest this transformation than to insist that the blues is embedded within an American anthem, both disrupting and fulfilling that anthem's ideals by troubling its rhythm?

Shockley produces further quirky disruptions of poetic and cultural time by punning alongside Black funk music. Her approach accomplishes a key effect of puns that Jonathan Culler finds exemplified in Shakespeare's plays:

> Punning frequently seems not so much the act of a character, expressive of an attitude, as a structural, connecting device that delineates action or explores the world, helping the plays (and also the sonnets) to offer the mind a sense and an experience of order that it does not master or comprehend. (8)

Through such disordering puns, Shockley's poem mimics the P-Funk song "Chocolate City," which was released when Washington DC was a city with a Black majority population and which urges Black residents not to be surprised "if [Muhammad] Ali is in the White House" and goes on to list Richard Pryor as the Minister of Education, Stevie Wonder as Secretary of the Fine Arts, and Aretha Franklin as the first lady. "post-white" offers its own subversive litany: "secretary of state frederick douglass who won't dread scott's claim of citizenship chief justice of the supreme court harriet jacobs master strategist speaker of the hour music moving mountains I'd let freedom ring with the harmonies of liberty." The run-on sentences whose lack of punctuation piles up syllables in nonconforming, uncertain relation have their effects expanded by the layering of meaning produced by punning allusions. Turning noun to verb, Douglass—in an alternative past leading to this alternative future—would with no anxiety accept that Dred Scott was a citizen, especially as Harriet Jacobs, fellow writer of slave narratives, would write the Supreme Court's majority opinion instead of Chief Justice Roger B. Taney who declared in that 1857 case that Black people "had no rights which the white man was bound to respect." Master of rhetorical and political strategies, Jacobs becomes "speaker of the hour," not just of the House and third in line for the presidency, but the person whose vision of justice is right on time. Shockley ends the poem by adapting James Brown's precursor funk tune "Papa's Got a Brand New Bag" and his "Say it loud! / I'm Black and I'm proud" to this new sense of (dis)order: "mama's got a brand new bag, and say it plain from that day forward we were all hip hop you don't stop being American." Changing genders and implying rather than stating the words of pride from Brown's song disrupts the expected closures of these phrases familiar to some of us. Yet, Brown's elided assertion of pride resides as more than residue in the knowing reader's mind as it is "plain" that Black pride makes hip hop's lack of stop, its ongoing variegated breakbeat, which consists of the DJ's formal transformation of a given song into its longest instrumental interlude, a possible articulation of the arranging life rhythms of those dwellers dreaming of their showers. Just as P-Funk knew, Shockley knows that the multiple rhythms of Black cultural life both sustain and puncture mainstream rhythms to reveal the self-defining Black being that was always there breaking the enclosure it was bound to and bound to be.

These gestures toward temporal expansiveness in Shockley's otherwise grounded social critique find even fuller fruition in Mackey's similar rhythmic punning and his distinctive anagrammatic play which are keyed to an even more expansive "we." Quoting Mackey's "Song of the Andoumboulou: 12" from *School of Udhra*, Andrew Mossin captures the terms of Mackey's broader and more fundamental dissent. Mossin writes, "Insofar as totalities are rejected in favor of contingent particularities, transcendence in favor of well-traversed limits, [Mackey's] poems likewise offer themselves as provisional, intentionally reticent accounts of 'Bound I. Insubordinate / Us' (*SU* 10)". Mossin emphasizes how Mackey adapts poststructuralist theories of the discursive constitution of the subject to his contrasting attention to gnostic and mystic traditions that seek foundationalist symbolic meaning in language. What bounds the "I" in discourse also somehow indicates how that "I" relates to the "insubordination" of possible mystic meanings, yoking seemingly opposed epistemologies into hints of other possibilities for knowing what "we" is. The fact that "Insubordinate" visually completes the line "Bound I." diminishes the full stop of the period after "I." Moreover, the "us" occupies one of what Lucy Alford calls Mackey's "hinged lines" by being one line down and justified to the right of page, seemingly hanging from the preceding line and "forming suspensions of time and attention" (Alford). The "I" and the "us" here are paradoxically "bound" by "insubordinate," are linked visually and syntactically to a dissent from bounded notions of selfhood to amplify that the "plot" of the poem validates how the "we" is "Breathing the smoke left by the gods' / exit." Whether abstracted into pronouns as linguistic constructions or presented as validated numinous reality in the Yoruba spiritual rituals described in the poem, the "we" circumscribes mystical potentiality as a limiting social construction while also internalizing—inhaling—whatever meanings the gods once had. Its yearning to fill that absence with itself may thus be all that the spirit is, as Derrida or Foucault might suggest, but that limited spirit nonetheless indicates not just absence but possible communion.

Mackey's "Poem for Don Cherry," which consists of "'mu' first part" and "'mu' second part," claims from its free jazz inspiration the sense that formal innovations compose each such "we" within, against, and across nations, traditions, and spiritualities through its intermittent, ongoing making. As scholars of Mackey have often noted, the Greek word "mu" indicates the punning etymological interconnection between "myth" and "mouth," unifying the act of speaking, the nonsemantic sound, and the spiritual stories that Mackey celebrates in Cherry's music. This etymology testifies in shorthand to a central aspect of Cherry that Mackey emulates, therefore, which is that, in Cherry's practice, Blackness becomes plenitude. In his albums "Mu First Part" and "Mu Second Part," Cherry's songs had become "explicitly global, sounding flutes from North America, the Middle East, and Asia; percussion from the Caribbean and India; a hunters harp from West Africa," and multitradition percussion instruments as well as "Carnatic singing, and rhythmic structures [alongside] pan-African vocalizations" (Roth 50). Moreover, Cherry "synthesized the musical and theologic practices of Buddhism, Sufism and Islam, Choctaw spirituality, Hinduism, Christianity, and a commitment to unconditional love"

(Roth 51). (It is worth noting that Mackey studied many of the same "Eastern" traditions.) These elements are combined in the hyper-improvisational structure of free jazz in which there is no central harmony or melody or standards against which to play as there was in bebop, so that, with or without his trumpet, Cherry could assert multiple myths simultaneously through his mouth. What he models for Mackey, then, is attention to constantly fluctuating aesthetic forms inviting but resisting the regular correspondences and closures of (Western) harmony, melody, progression, and resolution in favor of the disjunctions of parataxis and simultaneity.

The first part of Mackey's "Poem for Don Cherry" gestures toward this disruptive, unifying expansiveness in Mackey's typical "hinged" poetic structures. It opens, "The day before the / year begins woke to a / glimpse of her / scratchy legs." The next stanza describes "her" as "Lotusheaded, squatted," alluding to Lajja Guari, a Hindu goddess of fertility, a pregnant association for the day before the year begins. Of course, it is not the new year yet, so this is a more mundane day whose rhythmic and ritual significance nonetheless abides in the poem's last two "hinged" stanzas:

> Sat upside a Hill to
> wait for rain or
> watch the sun set,
>
> see
> how far the way we'd
> come
> went back

The jagged indentations amplify the breakdown of the roughly iambic rhythm of the first line of this passage. How does one "scan" the gap of the indentation whose possible anticipation of regularity in four one-syllable words is complicated by the next line, also of four one-syllable words, starting closer to the left margin than the first line of the stanza? Even if our tendency is to read past such spacing, that space nonetheless slows the eye, and certainly the "image" part of the "image-text" effects comprehension even if it does not get sounded in an out-loud reading. "Measure" here thus become both confine and possibility with the latter being indicated by each subtle visual and verbal variation that calls attention both to the pattern it could but does not fully complete and to other possible patterns toward which it gestures in its jagged (non)closure. Similarly, whatever ritual waiting for rain or for the sunset that occurs on whatever "Hill," the scratchy-legged "she" has become a "we" that seems to be simultaneously the two characters with which the section begins and a "we" much larger than a couple waking on New Year's Eve. Indeed, the visual scaffolding of the poem hints at the constantly renewing seasons aligned with communal rites of spring rather than only the winter of New Year's Eve in the northern hemisphere, with anticipation ("come") being the rightmost word, even as the poem emphasizes the retrospection of how far the "way we'd / come / went back." Is the poem a record of a couple's reflection at year's end in the name of a fertile future? Or is it enacting rituals of collective human yearning from which communal myth emerges to explain the mysteries of the past? Although seeking such distinctions makes sense, what seems to matter more is simultaneity, the syncopation of expected domestic rhythms with public ritual celebrations, local space and eternal time, rites in either case that sacralize quotidian life with the yearning for the spirit that may (or may not) already be manifest there in the yearning itself if not in the people.

I call this aspiration for meaning, this sacralization of yearning, the "human spirit" because Mackey's emphasis on its formal constitution in ritual practice—its status as a powerfully deployed discursive construct tied to genuine emotional need—acknowledges the complicity of "human" in the social death of colonization potentially to recuperate the concept as something possibly more genuinely "universal." In the "mu second part," likely the same "she"—her hairy legs reappear—becomes even more expansively holy: "And at the Stream, / who in cupped hands held it, / thirst, / or some worship, / whichever." Her holy hands hold "thirst" or "worship," "whichever," because neither serves to indicate fully the sense of loss at the heart of the yearning or the aspiration for fertility from potentially inert water that potentially empty gestures—her hands hold nothing tangible—want to produce from the fecundity of the "mud" and "eggs" and "likenesses." These are the kinds of correspondences by which gnostics perceive the larger meaning of existence outside of history, but Mackey warily and productively embodies those likenesses in situations that both affirm and question them. For example, these lines allude to "Preface to a Twenty Volume Suicide Note" by LeRoi Jones, a poet whom Mackey claims as a great influence, in which the speaker finds potential solace in how his daughter was "on her knees, peeking into / / her own clasped hands." Mackey adapts the girl's voiceless gesture, her moment of silent communion ultimately with herself that hints at Jones's antireligious skepticism, into a more faithful harbinger of sacral evocations available in quotidian human expression: "Noises / come out of it / Calls / *teo teo can*." That grammatically disjunctive call, referring to a singular source though "noises" is plural, directly quotes the nonsemantic voicing that Cherry offers in the recorded version of "Mu," a call that Mackey associates with "The coarse / cloth of Moorish cante." This juxtaposition suggests that one phrase translates the other or, more likely, as in free jazz improvisation, that they are irreducibly singular elaborations on shared rhythmic phrases, with the play of consonant sounds and their visual patterning insisting upon both likeness and difference. "Moorish cante" evokes a godly chant in flamenco as validation that such sound could be locus of whatever it is the clasped hands conjure. Ultimately, the conjuring comes from a source bearing a powerful likeness to what is usually meant by "the human," construing that burdened term as the possibly universal searching for the meaning of existence consistently insisting itself within and against whatever cultural norms have come to encrust it.

Mackey's reliance on the notion that form—the clasping of hands, the hinging of poetic lines—may *be* spirit or actually conjure it instead of only representing or constructing it only as discourse, clarifies the usefulness of "fugitive time" for comprehending such innovations. As Norman Finkelstein observes, Mackey is interested in the "unity of all rites," meaning the yearning and consequent pursuit of meaning essential to all human spiritual rituals, but Finkelstein does not note that what also unifies all rites, at least for Mackey, is a sense of time, of isolated moments of ecstasy generated during each enactment of the rites, as those moments resist the normal run of minutes and hours and associated social values. Finkelstein is right, then, that the "sacral" nature of the "rasp" of the *duende* in flamenco—its edgy aharmonious singing—exemplifies for Mackey the need to find spiritual possibility in moments of friction, moments Mackey describes as "beyond mastery." In other words, one may find manifestations of the spirit in a voice only when it is pushed during crucial, unsustainable moments beyond the normal realm of valued sound. Mackey's hinged lines and other disjunctive techniques seek to mimic such moments, which constitute Mackey's version of Fanon's declaration that "It is by going beyond [*en depassant*] the historical and instrumental given that I initiate my freedom ['que j'introduis le cycle de ma liberte']" (qtd. Omelsky 24). As Omelsky rightly reads Fanon, "refusing the past's determination of the present, [Fanon] pushes against that colonial, progressive conception of time" especially when one realizes that "Nonlinearity is built into Fanon's use of the noun *cycle*, which *Le Robert* defines not as a unidirectional movement or return but 'a sequence of constantly renewing phenomena'" (24). Mackey's "hinged" verse mimics Fanon's proposition that "a sequence of constantly renewing phenomena" momentarily and repeatedly intermingle spiritual transcendence with social liberation, the unifying aspirations of all rites with the discursive constitution and disruption of knowledge/power, in ways potentially more fully liberating than either alone.

"Alphabet of Ahtt," a poem "for Cecil Taylor," offers a key example of how Mackey points fugitive time toward the possible redemption of the "human." The poem takes its anagrams of "that" and "myth" to places of sound that refuse the rational sense similarly to how Taylor's Julliard-informed playing emphasized the piano as a rhythm rather than a melodic or harmonic instrument. According to Taylor, "Harmonic changes provided yesterday's dynamism" (qtd. Reed 157). Resisting "yesterday's" unities and progression of harmonic changes, the poem valorizes a "Scourge / of sound. Under its brunt / plugged ear unload." From the painful ("Scourge") rather than pleasing sound of Taylor's pounding fingers, orgasmic hearing spurts from ears, liberating hearing with as much blunt-force outflow as intake of sound. Expanding on such paradoxes through punning anagrams, the poem clarifies how the "we" of the poem, "Not / knowing why but reaching / elsewhere," found communion in such brutal, beautiful noise:

> Not without hope though
> how were we to take it as
> they yelled out, "Nathess's melismatic
> ttha"?

"They" assert the tongue-twisting name of some honored figure whose worship is characterized by a "melismatic" articulation—a single syllable sung in different tones in one breath, like *duende*—of a barely pronounceable anagram of "that." Is it the voice of a Muslim call to prayer? A Jewish cantor? An R&B singer? Generic difference matters here primarily as the measures—the techniques—of different cultural responses to similar hopes and as the varying instantiation of different unpronounceable names for that same yearning and its possible future fulfillment. Momentarily united by that constantly renewing cross-genre possibility, "we" danced to the music of this unknown melisma against the epistemology of a slavery-era spiritual: "'By and by,' / we would understand it someday" with "roundaboutness put aside," putting an end to "Tautologic / drift in which once more what spoke / of speaking spoke of speaking." As it is dubious that tautology could end tautology, it is the melisma that seems to matter, the momentary sound exceeding semantics and moving without progression or resolution, hinting at the spiritual and temporal "elsewhere" "they" and "we" yearn for separately but together, and that "they" and "we" therefore might, in that moment, actually inhabit beyond historical and instrumental givens.

This faithfully skeptical "we" as "sudden" as Shockley's leaves behind the "throngs" and the propertied subjects and the socially dead slaves for assertions of subjectivity potentially universal because unendingly, irreducibly, proleptically partial. The poem "ends,"

> anagrammatic
> ythm, anagrammatic myth …
> Autistic.
> Spat a bitter truth. Maybe misled but
> If so so be it. Palimpsestic
> stagger,
> anagrammatic
> scat

The intersection of rhythm and myth in the anagram "ythm," one of Mackey's most often cited, emphasizes once again that sound effect across time, whether standard "words" or not, constitutes meaning-making articulation. But because improvisation often works through repetition and revision, articulation may mislead if thought of only as linear, as without "roundaboutness." The structures of the lines in this stanza align "Palimpsestic" with "anagrammatic" and "stagger" with "scat," both conjunctions valorizing the nonlinearity and the intermittent layering of nonsemantic sound. In "scat," for example, there may be shared relation of human voices mimicking nonsemantic instrument play whose "stagger" potentially liberates from inarticulate isolation ("Autistic.") with its own inconsistent movement. Omelsky's characterization of Sun Ra's work thus seems apt for Mackey's too:

> This formal mode ["the proleptic inhabitation of this multifarious myth" in "accumulations" and "equations"] corresponds to the frequent, seemingly inscrutable silences, break, lulls, sustained pitches, and other iterations of openness that saturate Ra's poems and recordings, which together generate a sense of anticipation, gesture to imperceptible fugitive pathways, and evoke the openness of flight. (160)

Like Sun Ra's, Mackey's work expands its possibilities beyond the trenchant oppositions of West and Africa, future and present, knowledge and "roundaboutness," by paradoxically offering openness and anticipation as staggered artistic structures in which these oppositions are simultaneously present and thus less easily opposed. Silence and breaks structure flight through sound and its disruption, indicating an ecstasy that is available in this disjunction and repetition, singularity and recurrence, in and as our unsure and temporary togetherness in yearning not-knowing.

Here I must confess that the broader fugitive time of Mackey's approaches allowed me to notice more fully than I had before the innovative temporality of Shockley's "post-white." I have long read its last line only as an ontological claim about what it might mean to *be* American. But "whiteness" is certainly not over, so how can one *be* "post-white" except in aspiration? The poem's last line calls for its readers to "say it plain from *that* day forward," but what day could "that day" be? It is clearly distanced in time from "this day" on which I or you read the poem. Thus, one could read "that day" as a future day when one can genuinely be post-white, "the unanticipatable world of relief [from oppressive pain] that, impossibly, must be dreamed in advance" (25). Yet, given that the poem has emphasized music from the past, "that day" may also point toward a "post-white" moment that has somehow already preceded this current world of social death, almost as if the "whiteness" bearing its knees down on necks will have never been. After all, from that day, whatever day it was or is, "we *were* all hip hop you don't stop being American." Americanness, with its inevitable hip hop culture, apparently began with the middle passage and Jamestown instead or alongside of Plymouth Rock, with that original social death surely being erasure but also as surely being source. Here Shockley's "we" is more expansive than the exclusively Black "we" that "post-white's" title implies in its ostensible claim of only those Black "ones" who know the Blackness of torch songs and P-Funk's music. In our ever more fundamental and comprehensive and welcome critique of the pervasive workings of antiblackness, we cannot lose sight of each speculative "we" that Hartman narrates closely, for example, or in even the voiceless dead whose life rhythms Phillips tries to articulate by rearranging words, phrases, and even letters from the found language of court cases adjudicating to what extent Africans thrown overboard in a storm were insured property in her *Zong!* With these poets, we must continue to try to sense the dreams "we" have of ourselves, or at least consistently credit the forms that insist those dreams into existence.

Reed's observation about how the ecstatic operates in free jazz improvisation may overstate what Mackey and Shockley (and Phillip and Brand and Lewis) are doing, but not by much: "Black ecstasy is a site of liberatory risk or chance that does not posit a unique, lived experience of blackness but instead insists on blackness as a disruption of prevailing accounts of phenomenological lived experience" (153). I agree that, in the work of these poets, the most empowering implication of Blackness is its necessary epistemological and ontological disruptiveness. Blackness as rupturing senses of time and being: it is through this premise that poets imagine innovative form moving through colonial language to somewhere beyond the limits of racial discourse, beyond historical and instrumental givens, to instantiate, impossibly, the temporary utopia "we" might become in an instant. Such measures of fugitive time thus become ways for poetic structure to gesture toward possible existence outside of Western clock time, its enslaving capitalism, its encrusting racial discourses, the affiliated poetic metronome, and social death so that "we" can inhabit quite suddenly, if at all, the universal humanity we all will have been.

Works Cited

Alford, Lucy. "Suspension Bridges: The Poetics of Relation in Nathaniel Mackey's *Splay Anthem*." *MLN*, vol. 135, no. 5, December 2020, pp. 1258–282.

Culler, Jonathan. "The Call of the Phoneme." *On Puns: The Foundation of Letters*. Edited by Jonathan Culler. Oxford UP, 1988.

Finkelstein, Norman. "Nathaniel Mackey and the Unity of All Rites." *Contemporary Literature*, vol. 49, no. 1, Spring 2008, pp. 24–55.

Glissant, Eduoard. *Poetics of Relation*. U of Michigan P, 1997 (2010).

Hartman, Saidya. *Scenes of Subjection: Terror, Slavery, and Self-Making in Nineteenth-Century America*. Oxford UP, 1997.

Hartman, Saidya. *Wayward Lives, Beautiful Experiments: Intimate Histories of Social Upheaval*. W. W. Norton & Co., 2019.

Jackson, Virginia. "The Cadence of Consent: Frances Barton Gummere, Lyric Rhythm, and White Poetics." *Critical Rhythm: The Poetics of a Literary Life Form*. Edited by Ben Glaser and Jonathan Culler. Fordham UP, 2019, pp. 87–105.

Jackson, Virginia. *Dickinson's Misery: A Theory of Lyric Reading*. Princeton UP, 2005.

Mackey, Nathaniel. *Discrepant Engagement: Dissonance, Cross-Culturality and Experimental Writing*. Cambridge UP, 2009.

Mackey, Nathaniel. *Eroding Witness*. Selva Obscura P, 2018.

Mackey, Nathaniel. *Paracritical Hinge: Essays, Talks, Notes, Interviews*. U of Iowa P, 2018.

Mackey, Nathaniel. *School of Udhra*. City Lights Publishers, 1993.

Mossin, Andrew. "Unveiling Expectancy: Nathaniel Mackey, Robert Duncan, and the Formation of Discrepant Subjectivity." *Callaloo*, vol. 23, no. 2, spring 2000, pp. 538–62.

Moten, Fred. "The Case of Blackness." *Criticism*, vol. 50, no. 2, 2008, pp. 177–218.

Nelson, Angela M. S., ed. *"This is How We Flow": Rhythm in Black Cultures*. U of South Carolina P, 1999.

Omelsky, Matthew. *Fugitive Time: Global Aesthetics and the Black Beyond*. Duke UP, 2023.

Patterson, Orlando. *Slavery and Social Death: A Comparative Study*. Harvard UP, 2018.

Quashie, Kevin. *Black Aliveness, Or a Poetics of Being*. Duke UP, 2021.

Reed, Anthony. *Soundworks: Race, Sound and Poetry in Production*. Duke UP, 2023.

Roth, Paul N. "Teaching Jazz, Teaching Justice, and the Blackness of Don Cherry's Global Communion." *Jazz Education and Practice*, vol. 3, no. 1, 2022, pp. 48–58.

Sexton, Jared. "The Social Life of Social Death: On Afropessimism and Black Optimism." *Time, Temporality, and Violence in International Relations: (De)Fatalizing the Present, Forging Radical Alternatives*. Edited by Anna Agathangelou and Kyle Killian. Routledge, 2016, pp. 61–75.

Sharpe, Christina. *In the Wake: On Blackness and Being*. Duke UP, 2016.

Shockley, Evie. *the new black*. Wesleyan UP, 2012.

Shockley, Evie. "On Seeing and Reading the 'Nothing': Poetry and Blackness Visualized." *New Literary History*, vol. 50, no. 4, 2019, pp. 499–528.

Shockley, Evie. *Suddenly We*. Wesleyan UP, 2023.

Wilderson, Frank. "Introduction." *Introduction to Afropessimism*. liveright publishing, 2017.

9

Rhythm of the Brushstroke, Rhythm of the Body, and Cosmic Rhythm through Chinese Calligraphy and Painting

Yolaine Escande

We are governed by rhythms: the rhythms of the seasons, the alternation of day and night, our biological rhythms, starting with our heartbeat, and so on. These rhythms occupy a special place in Chinese civilization, for it has designed a universe never fixed, in perpetual motion, driven by the rhythm of cosmic breathing. This is particularly perceptible in the mountains, where mists, streams and clouds driven by winds, meteorological changes and earthquakes express the life of the universe as the Chinese conceive it.

The physical elements of nature are thus perceived as animated by forces, arteries or veins in which flows *qi*, that is, a vital breath identical to that animating living beings. In popular belief, the mountain is the body of the cosmic being: the rocks are its bones, the water its blood, the trees and grass its hair, and the clouds and mists the vapors of its breath. Consequently, cosmic breathing (*qi*) is the visible manifestation of the principle of life: the heart of nature, the mountain, beats to the rhythm of the universe.

Therefore, if the breath, *qi*, both universal and particular, animates all living things, ranging from human beings to plants, including constellations, rocks, clouds and animals, it cannot but also breathe "life and movement" (*shengdong*) into the works of art, such as paintings. This explains why one of the rules of composition in Chinese painting, on paper or silk with a soaked ink brush, consists of modeling the brushstroke on the temporal, rhythmic functioning of nature, in accordance with the principle of cosmic respiration, of "opening and closing" as expressed by Qing painter and theoretician Shen Zongqian (1721?–1803?), in his *Skiff on the Ocean of Painting* (*Jiezhou xuehua bian*, 1781):

> The genesis of heaven and earth can be described entirely in terms of alternating openings and closures. [In a vertical scroll], the lower part where the hand's

> activity begins corresponds to the opening, while the upper part where the painting unravels corresponds to the closing. [...] Just as a year is made up of an opening and a closing through the succession of the four seasons, a painting naturally consists of an opening and a closing through a beginning and an end.[1]

The "openings and closings" correspond to the respiration rhythm between the sky and the earth. The concrete realization of a painting, from the bottom to the top in a vertical scroll, particularly of "mountains and waters"—a phrase referring to the pictorial and literary landscape—resonates with the stroller's journey as they are climbing the mountain. The painter reenacts it while respecting a spatial and temporal progress on the scroll, itself modeled on the cosmic breath. In contrast, the European landscape painting is created through a process of accumulating successive layers, which is particularly important for the effects of light and darkness.

In the case of Chinese painting, rhythm plays such a critical role that it is the very first of the pictorial rules, considered the most significant, and formulated as early as in the fifth century. However, while rhythm may seem self-evident in the arts, whether in music, dance, or poetry,[2] in painting it seems to be more a question of color and tonality. This probably explains why Goethe and Kandinsky wrote their own color theories.[3] At any rate, rhythm does not concern color or tonality, but the brushstroke. Actually, rhythm implies regularity, similarity, cohesion, and, precisely, those elements are on what the Chinese painting brushstroke is grounded.

Although painting has existed in China since the Palaeolithic period, it was not recognized as an art and passed along as such until later, between the fifth and eighth centuries CE, and it was constantly compared to the art of writing. Even though Chinese historians knew that painting predated calligraphy, they constantly asserted that "calligraphy and painting share a common origin"

1 *Skiff on the Ocean of Painting* (*Jiezhou xuehua bian*, 1781), collected in Yu Jianhua (ed.), *Chinese Treatises on Painting by Categories* (*Zhongguo hualun leibian*, 1957), 2 vols., Peking, Renmin meishu chubanshe, republished 1977, vol. 2, p. 905. My translation.

2 For example, with regard to the Chinese arts, the *Book of Music* (*Yuejing*), one of the canonical books, starts with the following words: "The emergence of every musical tune has its origin in the human heart. The movement of the heart is produced by what exists: this movement, felt in the presence of what exists, takes shape in sounds. When these sounds are responded to by others, a variation is born. If these variations are combined in rhythm, we obtain a musical tune " (*Book of Rites*, beginning of chapter 17, Yueji 樂記). 凡音之起由人心生也。人心之動物使之然也。感於物而動故形於聲。聲相應故生變變成方謂之音。

3 Johann Wolfgang von Goethe (1749–1832), *Le Traité des couleurs*; Vassily Kandinsky (1866–1944), *Du spirituel dans l'art*.

(*shuhua tongyuan*) with regard to pictorial art, to give greater importance to the latter. In this expression, *shu* 書 means both "writing" and "the art of writing." Chinese writing appeared around the sixteenth century BCE, calligraphy as an art of writing in the first century CE.

The definition of the term "painting" in Chinese stems from this concept: it has nothing to do with applying colors. The etymology of "painting", *hua* 畫, is in fact a hand (手 or 巾) holding a brush (聿), above a cultivated field (田). Originally, *hua* meant "to draw, to outline a plan." Later, it came to designate the act of executing a painting and the product of this activity.[4] As a result, *hua* expresses the act of painting and its result, namely the brushstroke. This is why painting in Chinese is unrelated to the fact of coloring, as the painting theorist Zhang Yanyuan (ninth century) explains (trans. Bush and Shih 62).

The Six Rules of Chinese Painting

This also justifies the importance of the first two rules of Chinese painting that apply to painting and calligraphic art alike. In fact, there are six rules in Chinese painting, the first of which is: "the breath resonance gives life and movement," and the second: "the bone method in the use of the brush," or "structural method in handling the brush." The first may also be translated as "the breath resonance gives the movement of life" or "the breath rhythm generates the movement." These rules were laid down by a theoretician from the fifth century, who was also a painter at the Liu Song court. This treatise has remained a reference throughout Chinese history right up to the present day, and these eminent rules, particularly the first, have been interpreted differently across time. The fact remains that, for several centuries, these first two rules have been the bedrock of the scholarly theory of painting.

The six rules stated by Xie He (active around 479–502) are the following:

> Although there are six rules in painting, few painters have been able to master them all and, from antiquity to the present day, each has excelled in only one of them. What are these six rules? The first is the resonance of the breaths that

[4] See the paragraph "On the origins of painting" in Zhang: "The *Theory of Primitive Spellings* (*Shuowen*) [states]: '*Hua* means "to limit, to delimit"; it represents the diked path that separates fields, which is why it is a stroke (*hua*)'" (28).

> gives life and movement. The second is the method of the bone in the handling of the brush. The third is fidelity to the object in depicting the forms. The fourth is conformity to genre in the application of color. The fifth is the appropriate arrangement in the composition. The sixth is transmission through copying. (trans. Bush and Shih 39–40)

The expression "the resonance of breaths" or "breath resonance" (*qiyun*) seems to match to the notion of rhythm quite accurately. A literary treatise of the utmost importance in the history of Chinese literature and a contemporary of Xie He's, Liu Xie's (*c.* 465–522) *Literary Mind and Carving of Dragons* (*Wenxin diaolong*), cogently explains that in poetry if *qi* is the vital creative force, for *qi* to be successfully expressed in a given composition, words and phrases must be arranged harmoniously. *Qi* may be positive or negative; its harmonious and rhythmic arrangement is necessary to achieve an artistic composition. In chapter 33, entitled "Musicality" (Shenglü), Liu Xie gives the traditional definition of *yun* and its connection to breath:

> The force of breath (*qi*) is fully expressed in harmony (*he*) and rhyme (*yun*). *He* means the harmony of different sounds and tones. When two final vowels have the same sound, it is *yun*. (trans. Shih 355)

Yun 韻 is a picto-phonogram made up of two elements, one semantic 音 meaning "the sound," the other phonetic 員. The narrow meaning of *yun*, according to Liu Xie, is "rhyme." In this, he clarifies the definition in the first etymological dictionary, the *Shuowen jiezi*, dating from the end of the first century CE:

> *Yun*: harmony (*he*). [...] Formerly, identical to "equal"" (*jun*). (Xu 58)

The rhyme of the breath or breaths is rhythm, in poetry like in painting. Xie He's first rule deals with the "vitality" of the represented subject (the *qi*), the "harmonious" aspect of the execution (the *yun*), and the "liveliness" aspect of the execution (the visual result), elicited from the method advocated in the second rule, that is, the calligraphic brushstroke. Vitality, arranged in rhyme, is that not rhythm?

In practical terms, what does this rule correspond to? Let us turn to the original handmade manuscripts by artists known to Liu Xie and Xie He, such as an autograph by the poet Lu Ji (261–303) (Figure 9.1).

This calligraphy can be used to illustrate what *yun*[5] represents: the suppleness in the use of the inkbrush is perceptible, the harmony of forms, regular but not

[5] Soper (419) and Fong (159–60) explain that the first meaning of *yun* is "rhyme," and that it later takes on the meaning of "rhythm" and "poetry."

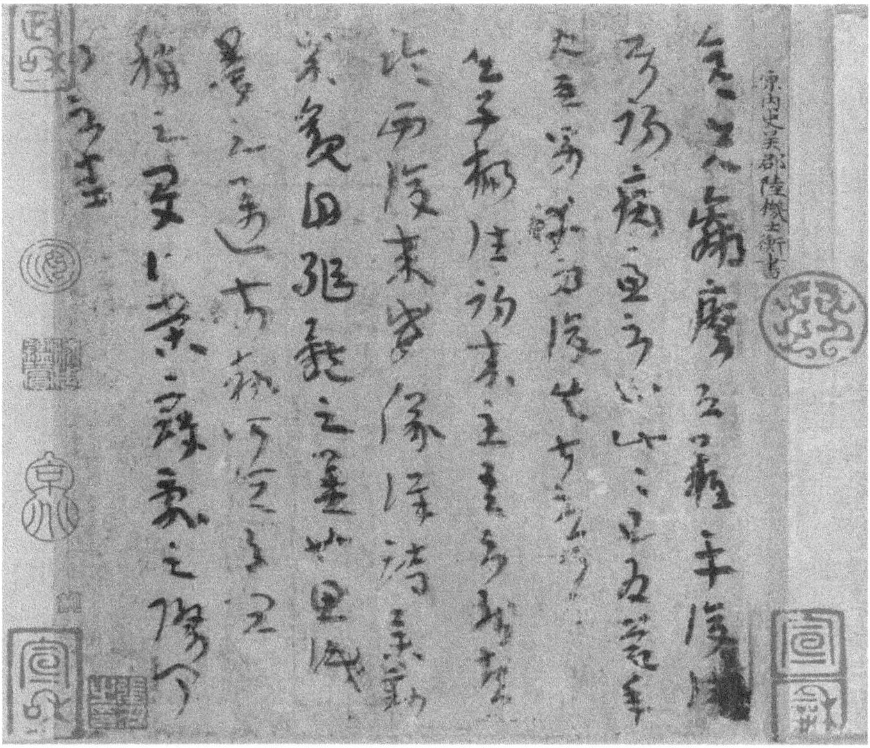

Figure 9.1 Lu Ji (261–303), *Letter (Pingfu tie)*, ink on paper, 23.7 × 20.6 cm, Beijing, Ancient Palace Museum. Public domain.

stereotyped, in motion and rhythmic, embodies the effect mentioned in the etymology of the term.

In painting, at the time of Xie He, "breath resonance" applies to representation, particularly in portraits; in calligraphy, it pertains to the visual result of the brushstroke. The second rule is borrowed directly from the calligraphic method, that of the "bone"; it consists in tracing a stroke full of dynamism, that is, marked by an inception, a development, and an ending. The third and fourth rules deal with conformity to form and genre, which can be understood as respect for the theme being dealt with, but also realization according to the medium: silk, paper, or lime-coated wall. The fifth concerns composition, that is, the relationship between the various calligraphic or pictorial elements. As for the sixth, it is the very condition for the existence of art, based on learning and the need for conservation through copying.

In his treatise, Xie He ties brushstroke to vital energy (*qi*), harmonious composition capacity, and resonance (*yun*). In his time, calligraphy was already an institutionalized art, having produced its own masterpieces. Consequently, Xie He insists on the calligraphic quality of the brushstroke, which must convey the represented object's vital energy.[6]

A few centuries later, theoretician Zhang Yanyuan (ninth century), who authored the famous *Record of Famous Painters Through the Successive Dynasties* (*Lidai minghua ji*, 847), started his treatise presenting the six rules of painting formulated by Xie He and then commented on them:

> I, Yanyuan, believe that some ancient painters were able to go beyond formal resemblance because they gave priority to the breath of bone structure (*guqi*). What they were looking for went beyond formal resemblance. It is difficult to discuss this with ordinary people. (Zhang 32)

For Zhang Yanyuan, it is clear that the brushstroke is fundamental: the structure of the scroll is constituted by its composition, and by the stroke itself. At that time, breath circulates within these lines of forces themselves, particularly in the "mountains and waters" paintings, in full bloom at the time. The first criterion whereby painters are appraised is therefore clearly stated: their brushstroke and in particular their "bone's breath" (*guqi*) or "bone energy." Zhang continues:

> Today's painters, when they succeed in obtaining formal resemblance, do not give rise to a resonance of breaths. On the contrary, if they were to seek the resonance of breaths in their paintings, the formal resemblance would appear [quite naturally]. (32, my translation)

Zhang Yanyuan does not neglect formal resemblance, which here refers to the ability to replicate the models of the ancients (Zhang 32). But this should not take precedence. The artist must seek for the internal and harmonic breath (the breath resonance) and convey it in their brushstrokes. This entails that resemblance follows, naturally. Therefore, Zhang Yanyuan upends the considerations expressed by previous theoreticians: they indeed prioritized formal resemblance, which implied the transmission of artworks by copying, being the sixth rule of painting.

What does Zhang Yanyuan's "bone energy" mean? He proceeds:

> Bone energy and formal resemblance are both rooted in the firm intention and derive from the use of the brush. That is why skilled painters also excel at calligraphy. (Zhang 32, my translation)

6 For more details, see Xie He, *Evaluation of Ancient Painters* (*Guhuapin lu*), trans. Bush and Shih 39–40.

Thus, there is no doubt that what is at stake is the use of calligraphic technique in painting. However, this interpretation remains limited to the animated effect of the stroke, and it does not summon the skillfulness of the calligrapher, contrary to what prevails in calligraphic appreciation: in calligraphy's theory, the quality of the brushstroke is explicitly related to the moral or spiritual quality of the calligrapher. This introduction of the evolution of the first rule of painting is indeed necessary to understand that the preeminence of the brushstroke in Chinese painting, and its appreciation, should not be taken for granted. In other words, in painting, there is no immediate representation of the brushstroke as rhythm of the world.

The Literati's Conception of *qiyun*—Rhythm of the World

In Chinese pictorial theory, what matters is, essentially, the vital energy's circulation in the painting, like in calligraphy, through the brushstroke: but does it correspond to that of the represented object in painting, to that of the brushstroke itself in calligraphy, or to that of the artist at work? Do the brush holders refrain themselves from conveying the rhythmic energy animating the universe, particularized in one object—the landscape—thus removing themselves from their gesture? Or do they harness their own vital energy, making use of the landscape as an excuse? Or do they deliver the result of the relationship they have established with the object? Depending on the period, the theory evolved, and the meaning of breath slightly shifted.

Hence, during the eleventh century, theoretician Guo Ruoxu in his *Notes on what I have Seen and Heard in Painting* (*Tuhua jianwen zhi*) offers a definition of what can be called the literati's art:

> Xie He stated: "The first [of the six rules in painting] is called breath resonance that gives life and movement; the second, structural method in use of the brush; the third, fidelity to the object in portraying forms; the fourth, conformity to kind in applying colors; the fifth, proper arrangement in composition; the sixth, transmission through copying."
>
> This essential definition of the six rules has not been altered throughout antiquity. Now the last five, from "structural method in use of the brush" on, are open to study. However, "breath resonance" (*qiyun*) necessarily involves an innate knowledge. It certainly cannot be secured through cleverness or close application, nor will time aid its attainment. It is an unspoken accord, a spiritual communion, something that happens without one's knowing how.

> I have already attempted to discuss this subject, and I venture to note that the extraordinary works of the past were mainly those by talented worthies of high position or superior gentlemen in retirement, who cleaved to benevolence and sought enjoyment in the arts or explored the abstruse and plumbed the depths, and their lofty and refined emotions were all lodged in painting. If a person's spiritual value has been high, his breath resonance cannot but be lofty. If breath resonance is already lofty, life and movement [in his works] cannot but be achieved. As it has been said, in the highest heights of the spiritual, it can deal with the quintessence. (trans. Bush and Shih 95–6, slightly modified)

For Guo Ruoxu, artists belong to an elite, animated with "breath resonance" (*qiyun*), thus endowed with moral and spiritual values, as a prerequisite but no sufficient condition for their works to be remarkable. In his interpretation of the first pictorial rule, *yun*, "harmony, resonance," qualifies *qi*, "breath." *Qi* embodies vitality, the activity of the artist's heart, whereas *yun* is its resonance, the harmonious vibration. Thus, *qiyun* is altogether inextricably bound up to both the artist and the artwork, taken as a whole. The artwork is then the coincidence of the heart and its outward expression. The "movement of life" is the result of a spiritual attitude, visible in the artwork.

Accordingly, *qiyun*'s meaning covers two levels of reality: it refers both to the painter and to his or her human quality, but also to the result of the painted work, that is, its superior quality of life movement. Guo Ruoxu's theory differs from Zhang Yanyuan's: unlike Zhang who considered that *qiyun* concerned only living beings and not inanimate objects, Guo relates *qiyun* to all paintings, considered globally as the accomplishment of an artist, of their personality. This points out to a remarkable evolution. Until then, calligraphy had been considered the direct expression of a character, notably on occasion of the imperial examinations,[7] but in painting, it was by no means the case. Following Guo Ruoxu, painting acquired the same status as calligraphy, at least theoretically.

It is no coincidence that precisely at that time, under the Song dynasty, the literati's artworks started to be signed, that they set inscriptions on their paintings and that their art started to flourish. In other words, from the eleventh century onwards, the medium conveying the rhythm of the universe, through the painter, is the rhythmic stroke traced with the inkbrush.

7 The imperial examinations' purpose was to select officials depending on their capacities, believed to be visible in their calligraphic expression. See Escande 148–69.

The task of rendering the vital breath of the cosmic body is then explicitly entrusted to the pictorial brushstroke, tracing the "lines of forces" composing a scroll. The "mountains and waters" painting is ranked first in the pictorial hierarchy as early as the tenth century, and it expanded remarkably during the fourteenth century. Wang Meng's famous painting *Living in Juqu Forests* (Figure 9.2) is particularly illuminating regarding the rhythmic strokes and the expression of the landscape's "lines of forces."

Figure 9.2 Wang Meng (1308–1385), *Living in the Juqu forests (Juqu lin wu)*, 1378, vertical scroll, ink on paper, 68.7 × 42.5 cm, Taipei, National Palace Museum. Public domain.

This vertical scroll is striking for its density and abundance, and for the absence of any horizon or perspective; the eye pauses on the lower part of the scroll where breathing begins, revealing a stream on which a skiff sails, a few trees and a succession of small inhabited huts sitting on the high points at the center of the scroll. The eye follows the movements of the trees and rocks, which seem to breathe thanks to the visible presence of the "arteries of the dragon" or the landscape's lines of forces, which organize the work from the rocks at the bottom left to the inscription bearing the title of the painting at the top right. The shimmering colors of the foliage, complementing the brushwork effect of numerous short, dry, sharp strokes, create the sensation of a rhythmic regularity, of breathing, as much in the ripples of the watercourse as in the tree trunks, the foliage, or the odd rocks.

Indeed, painting is *a priori* addressed to scholars, capable of reconstructing, in their minds, the path of the brush on the scroll. In François Cheng's translation, the painter and theoretician Shen Zongqian (1721?–before 1803) states:

> The brushstroke is the backbone of painting. Trees and rocks have shapes of infinite variety; they are born under the artist's brush as he, guided by his spirit, draws strokes. These strokes, in turn, guide the artist's spirit. There is thus a reciprocal relationship between the artist's spirit and the traced strokes. In a painting, all the traced strokes, born of necessity and charged with flavor, form the skeleton without which muscles and tendons would not connect, flesh and blood would not come to life. This is why the Ancients placed so much emphasis on the art of the brushstroke. (trans. Cheng 41, my translation to English)

The word that F. Cheng translates as "spirit" is the Chinese word for "intention" (*yi* 意) or intentionality. The brushstroke is what embodies the artist's intention on the one hand, and on the other, what makes the beholder not a mere external spectator but an actor in their own right by reliving the gesture internally. What is at stake is not the relationship between the painting-object and the scene represented, but rather the relationship forged between the painter and the person reliving their experience.

The Rhythm of the Brushstroke

Chinese calligraphy, an art deploying itself on a bidimensional space, faces severe constraints, which pertain to the writing of the character, the number of strokes and the order of their tracing, the chosen style, the relationships between ink and brush, ink and paper, brushstroke and paper, and so on. The calligraphic language, which is *the* reference for the Chinese painters, is molded by these constraints. At the same time, calligraphy stems from the interactions between the writer's operation and the material, and the back and forth movements of the writer's body to and from the paper. When the calligrapher presses the inkbrush on the paper, the "heart"—which is both the calligrapher's and the brush's—reacts depending on the quantity of ink, the firmness or the suppleness of the brush hair, the thinness or the hardness of the paper, on the thickness or the fluidity of the ink, and so on. Conversely, the paper reacts to the writer's hand and body, thus influencing or conditioning them, to which the latter respond in turn. Therefore, the whole body is modeled by the brushstroke. The assimilated calligraphic practice can be found in painting's composition and brushstroke too.

Moreover, the brushstroke, *hua*—the same term used for painting in Chinese—is not a line in the sense given by Merleau-Ponty to Klee's line. In *L'Œil et l'esprit* (*The Eye and the Mind*), Merleau-Ponty states: "There are no lines visible in themselves" (73), but he claims that the line marks a distinction between this thing and another thing (71–4). He goes on to discuss the importance of color in Matisse and Klee (74), because it is color that "makes things visible" (74). In Chinese painting and calligraphy, the stroke is not a line that renders things visible. Nor is it a surface, or else it would be both line and surface; it has no contour that would then need to be filled. The brushstroke is above all a dynamic *body*, which is why it must exist in itself, have an existence "as marvellous as that of nature" (Sun Guoting 2:269, my translation), in other words, be endowed with the same existential value as a blade of grass, a cloud, or a rock. In the calligraphic brushstroke, there exist very precise techniques allowing for the impression of life in a stroke, to provide it with a dynamic body, by, for instance, applying more ink at the beginning and at the end of a stroke, so as to inject life to a group of strokes or a pictorial composition. Yet, on principle, one traces the strokes by applying a regular rhythm.

In other words, when we look at a stroke, it is altogether a plastic sign, because it does not mean anything per se, it has no sense (except the horizontal stroke meaning "one") but that which we interpret arbitrarily from its shape. There is no contour in the brushstroke: it is drawn from the outset with the fullness of a surface, while being that which slits through and divides the space, according to a Western geometric vocabulary, or that which unites and connects the different elements in the space in the Chinese vision. For the Chinese beholder, the stroke is first and foremost dynamism whose experience is revived intimately, in their body. The brushstrokes are not perceived as lines of separations or boundaries between the strokes and the surface of the paper (Pelletier et al. 251–70), but as "rhythmic movement" strokes, interrelated in an organic relation.

That is why, although the appreciation may seem to concern the brushstroke, it actually revolves around the relation between the beholder and the artist, and overall, the beholder and the universe's rhythmic movement through the artist. Since the eleventh century, the Chinese tradition has conceived that the beholder of a painting, when observing a brushstroke on a scroll, not only triggers a mental representation of the gesture producing the movement of the brush, but also perceives or feels the brushstroke in their own body, precisely like in the calligraphic tradition. In this process, the rhythm of tracing is essential, all the more so as the brushstroke cannot be reworked, redressed, let alone wiped out. Therefore, in the apprenticeship process, copying a model does not consist only in imitating forms, but in internally reliving the brushstroke through re-experiencing the gesture and through its actualization. In other words, through the brushstroke, the calligrapher or the painter beholding it establishes a bodily bond with the writer. Incidentally, the calligraphic and pictorial vocabulary newer mentions the "beauty" of a brushstroke, but its "bone, sinew, blood, and flesh" (see Escande 155–81). Thus, when reliving the corporeal gesture leading to the brushstroke, by seeing its trace or its imprint, what is at stake is the bodily perception of the rhythm, objective or indexical analysis is by no means relevant.[8] On the contrary, what is recorded is the bodily perception of a rhythm. Nor does this process call for fictional "immersion" as it might be imagined when reading of a story,[9] which presupposes an opposition or contrast between perception and imaginative activity, whereas in the present case it appears there is no disconnection from reality when reliving the corporeal gesture.

The rhythm of life is not only reproduced in the way the stroke is traced, but also in the scroll's composition. It is the case in Huang Gongwang's (1269–1354) painting *Living in Mounts Fuchun*, dated 1350 (Figure 9.3), one of the

8 Objective or indexical study naturally creates a distance between the person examining the trace and the person who produced it.

9 In *Pourquoi la fiction?* (pp. 182–87), Jean-Marie Schaeffer defines what he calls "fictional immersion" in terms of five characteristics: state of imaginative activation, the doubling of worlds, a dynamic state, a state of affective investment, and the absence of erroneous beliefs.

Figure 9.3 Huang Gongwang (1269–1354), *Inhabiting the Fuchun Mountains*, 1347–1351, ink on paper, 33 × 636.9 cm, Taipei, National Palace Museum. Public domain.

masterpieces of Chinese painting, serving as a study model to impart the pictorial principles. In the structure of this long horizontal scroll, the beholder can sense the continuity spanning from right to left, giving breath to the painting; this is due to the dynamic structuring lines that provide rhythm to the composition. From this point of view, in such artworks, stretching in their width, the formal elements tend to develop horizontally, favoring an impression of stability, which may easily prove monotonous.

But Huang is exceptionally talented for offering rhythmic vitality to his painting: he alternates close and distant views, vertical shapes (mountains, trees) and horizontal elements (water, riverbanks, distant hills), empty and filled spaces. Because of the rhythm of the alternations following the principles of Chinese pictorial composition—open/closed, dense/scarce, full/empty, dark/light, far/close, high/low—Huang gives birth to a living breath: at the beginning, on the right-hand side, one sees a group of hills, then a large stretch of water develops, recognizable given that the paper was left blank, finally a solitary mountain arises and revives a vertical shape before leaving distant hills "tailing off" in the background.

As a matter of fact, a painting is an experience, especially after hiking in the mountains, and it can be reproduced in a painting, as the painter and scholar Leng Qian (active in the fourteenth century) asserts in an inscription on his painting of *Mount Boyue* (Figure 9.4): "I retain [this journey] in my heart and body, which has captured my thoughts and my spirit" (留我心身, 遏我神思). The beholder can replicate this experience, reconstructing it in his or her mind based on the rhythm of the brushstroke.

This is one of the earliest surviving paintings of a journey in the history of Chinese art. It is mentioned on the scroll itself, by the painter himself, who is both the traveler and the beholder.

The painter, Leng Qian, was a Taoist master and musician of the Yuan, a close friend of statesman Liu Ji's (Little and Eichman 372). The painting is on paper, in rather dry ink. The theme is a journey to the Yellow Mountains in Anhui undertaken by Leng Qian and Liu Ji in 1343, each writing an inscription on the painting. Leng Qian's inscription, the longest on the right-hand side, is in regular script and reads:

> Onward we went in successive stages for several tens of *li*, and we began to catch sight of the Celestial Gate [a famous cliff in the Yellow Mountains]. We paid homage to the sacred images and gazed with enjoyment at mountains and rivers. … Far-off peaks rose, one above the other, piercing the clouds and reaching the heavens. Forests groves spread out before them. I asked one of the locals about [this view], and he answered: the Yellow Mountains. It could be reached in a day's journey, so we set off upon the trek. Along the way we encountered strange

Figure 9.4 Leng Qian (active in the fourteenth century), *Mount Boyue* (*Boyueshan*), dated 1343, ink on paper, 84.4 × 41.4 cm, Taipei, National Palace Museum. Public domain.

> pines and wondrous rocks, majestic peaks and sheer cliffs, flying waterfalls and bubbling streams, cinnabar terraces and [illegible character], emaciated monks and strange animals, the cries of gibbons and the chirping of birds.
>
> I retain [this journey] in my heart and my body, which captured my thoughts and my spirit. Creeping vines led us to the summit of the mountain. The Blue River stretched out in a ribbon and Jinling [Nanjing] appeared clearly. From north to south, there were no limits; in all directions, the immensity stretched out, [as if] displayed on a screen before my eyes. Liu said: "We entered the mountains of the immortals by mistake." We never ceased to exclaim. So I tried to depict *(moxie)* this and to apply myself by daubing. Created by the Taoist with the yellow turban, Leng Qian. (trans. Little in Little and Eichman, slightly modified)[10]

Leng Qian points out the importance of this journey, he remembers in his *heart* and in his *body*, because he has physically climbed up the mountain. He experienced the spatial and temporal immensity as he reached the summit, which his friend calls the realm of immortals. However, he does not paint the summit, or the view from the top of the mountain.

What is worth noting here is that the journey leads to the paradise of the immortals, and that *the painter can remain there physically and in spirit because he paints it*. The painting we behold on a scroll does not simply mean he turns the land into a landscape, like in the European tradition (see Clark). Much more profoundly, he *becomes* immortal, because precisely, he can sense this unlimited space and time, and merge into the world surrounding him.

And, through the painting, Leng Qian allows the beholder to relive such an experience, by wandering within the painting. The aim is not to stop at the pleasure of the eye, but to incorporate oneself into the landscape and become one with it. This bodily integration is made possible through the experience of hiking first, then through the brushstroke, by means of the rhythm. The experience of keeping this journey "in my heart and in my body," which "captured my thoughts and my spirit" is replicated for the beholder, when they reconstruct it in their mind, thanks to the rhythm of the brushstroke.

Indeed, the bodily experience of the journey in the mountains, experienced by the painter and traced in the shape of rhythmic brushstrokes, is conveyed to the beholder. The rhythmic course of the mountain hike allows the traveler to stroll following the rhythm of the universe's changes, particularly perceptible

[10] Little's translation is not complete.

in the mountains, and in doing so, *Chères et chers amis, chères et chers collègues*, to participate to them actively. The painter and hiker render this cosmic respiration through the breath momentum of the brushstroke, and the beholder can in turn sense it and relive it when visually and bodily roaming along in the scroll, by making use of their own memories of hiking in the mountains.

And through painting, the painter offers the viewer the chance to relive such an experience through a visual journey through the work. The aim is not to stop at the pleasure of the eye, but to enter into the landscape and become one with it. This "*corps à corps*" is achieved first through the experience of walking, through the rhythm of walking and breathing, and then through the rhythm of the stroke that takes it in and gives it back.

Thus, by making use of imaginative activity, and thanks to their mastery in brushstrokes, when scanning the pictorial rhythmic strokes, the beholder of the painting mobilizes and meshes the relived gesture of the brushstroke and the memory of the mountain hike marked by the changes of the universe's tempo.

Works Cited

Bush, Susan, and Hsio-yen Shih. *Early Chinese Texts on Painting*. Harvard UP, 1985.

Cheng, François. *Souffle-Esprit. Textes théoriques chinois sur l'art pictural*. Seuil, 1989.

Clark, Kenneth. *Landscape into Art*. John Murray, 1949.

Escande, Yolaine. "Tang Dynasty Aesthetic Criteria: Zhang Huaiguan's *Shuduan*." *Journal of Chinese Philosophy*, vol. 41, nos. 1–2, 2014, pp. 148–69.

Escande, Yolaine. "Some Aesthetic and Artistic Categories in Chinese Painting and Calligraphy." *The Bloomsbury Research Handbook of Chinese Aesthetics and Philosophy of Art*. Edited by Marcello Ghilardi and Hans-Georg Moeller. Bloomsbury Publishing, 2021, pp. 155–81.

Fong, Wen."Ch'i-yun-sheng-tung: Vitality, Harmonious Manner and Aliveness." *Oriental Art*, vol. 12, no. 3, 1966, pp. 159–64.

Goethe, Johann Wolfgang von (1749–1832). *Le Traité des couleurs* [1808–10]. French translation by Henriette Bideau, accompanied by three theoretical essays by Goethe; introduction and notes by Rudolf Steiner. Triades, 1973.

Kandinsky, Vassily (1866–1944). *Du spirituel dans l'art, et dans la peinture en particulier* [1911]. French translation by Pierre Volbout.... Denoël, 1969.

Little, Stephen, and Shawn Eichman. *Taoism and the Arts of China*. U of California P, 2000.

Merleau-Ponty, Maurice. *L'Œil et l'esprit*. Gallimard, 1960.

Pelletier, Jérôme, et al. "Brushstrokes with Emotion." *Inter-Culturality and Philosophic Discourse*. Edited by Yolaine Escande, Vincent Shen, and Chenyang Li. Cambridge Scholars Publishing, 2013, pp. 251–70.

Schaeffer, Jean-Marie. *Pourquoi la fiction?* Seuil, 1999.

Shen, Zongqian (1721?–1803?). *Skiff on the Ocean of Painting* (*Jiezhou xuehua bian*, 1781). *Chinese Treatises on Painting by Categories* (*Zhongguo hualun leibian*, 1957). Edited by Yu Jianhua. Renmin meishu chubanshe, vol. 2, pp. 865–911.

Shih, Yu-chung Vincent. *The Literary Mind and the Carving of Dragons*. The Chinese UP, 1983.

Soper, Alexander C. "The First Two Laws of Hsieh Ho." *The Far-Eastern Quarterly*, vol. 8, no. 4, 1949, pp. 412–23.

Sun Guoting (seventh century). *Treatise on Calligraphy* (*Shupu*, 687). *Chinese Treatises on Painting and Calligraphy* (*Lidai shufa lunwen xuan*). Edited by Huang Jian. Shuhua Chubanshe, 1979, pp. 124–32.

Xu, Shen (30–124). *Shuowen jiezi* (lit. *Theory of primitive spellings and explanations of derived forms*, 100). Zhongghua shuju, 1987.

Yu, Jianhua (ed.). *Chinese Treatises on Painting by Categories* (*Zhongguo hualun leibian*, 1957). Renmin meishu chubanshe, republished 1977, 2 vols.

Zhang, Yanyuan (ninth century). *Record of Famous Painters Through the Successive Dynasties* (*Lidai minghua ji*). *Chinese Treatises on Painting by Categories* (*Zhongguo hualun leibian*). Edited by Yu Jianhua. Renmin meishu chubanshe, 1957, vol. 1, pp. 27–40.

10

Reading Rhythm as Translation: Bei Dao's "The Answer"

Nick Admussen

Rhythm is contextual.[1] It is learned (or "entrained").[2] It almost always appears in the form of a layered polyrhythm, it requires subjective intepretation, and it appears and disappears not exclusively due to effects in the physical world, but also due to the presence or absence of those that can receive it.[3] Rhythm is an interaction between a *langue* of temporal patterns, held by listeners, and the *parole* of a specific shape in time.[4] Rhythm is semantic: it carries ideas, associations and affects flexibly, and there are few, if any, deterministic relationships between a specific rhythm and the human experience it provokes.[5] Therefore, formalist

1 This essay was provoked by reading Handel and Lawson, who performed a psychological experiment on undergraduates in which they were encouraged to tap along with complex polyrhythms. Their responses—which rhythms they reproduced and how—were highly dependent on nonrhythmic qualities of the beat (its tone, its volume), and identical interpretations differed strongly between individuals. Even further, though, we know that all waves must move through a physical medium, and that different media affect waves differently; all sound is rhythmic and all sound is contingent upon context.

2 London (6) describes the importance of cultural training in creating listeners for meters and other rhythms: "these entrainment behaviors are highly practiced: from early childhood we are steeped in a musical environment, and have many opportunities to develop and hone our attentional habits ... when we attend to a piece of music, we are rarely starting from metric first principles."

3 Handel and Lawson (103): "rhythmic organization exists at all levels of activity. What musicians, carpenters, and dancers have in common is an appreciation of rhythmic levels." For the dancer, the music has one rhythm, gravity has another, the body's endurance has a third, and fellow dancers have a fourth; there are many more. Note the importance here of *appreciation*, discernment, judgment; it is this habit that divides foreground rhythm from background noise, which is also likely rhythmic if appreciated to be so.

4 Adapted from Saussy (47–9), in which he advances the idea that folklore is a *langue* in Saussure's sense, and an individual folktale is a *parole*.

5 Examples of this abound. Early Communists used Mao Zedong's regulated, metrical, classical-style poetry to argue to military and political elites that he was a virtuous and powerful leader. This process contributed to the canonization of five-character and seven-character meters as orthodox Socialist rhythms, even though they were absolutely central to imperial and feudal structures of power. Over the course of perhaps fifty years, these very old poetic rhythms saw their associated ideologies, affects, audiences, and meanings radically transformed. See Yang.

literary analysis, especially the analysis of rhythm in poetry cannot be reduced to the notation of scansion or the binary of beat and rest. This is because social forms, literary forms, and material forms interact, interfere, and intermingle.[6] The translation of rhythm requires abandoning totalizing concepts and instead partially interpreting an infinitely thick,[7] layered set of cultural interactions; although this essay will attempt to comprehensively translate a poetic rhythm from one language, place and (most importantly) time to another, it will remain necessarily partial.[8]

Imagine, by way of example, a person playing a hand drum in a public park. One rhythm is the beat that the drummer intends to carry. A second is the pattern of mistake—a little early sometimes, a little late others—that identifies the drummer as a specific human being. An experienced listener would be able to hear both rhythms at once, the ideal rhythm and the actual rhythm, and what we think of as the beat would be the polyrhythm that includes both signaled intent and actual practice. But drumming in this particular city park is traditionally done in groups, and this drummer might be signaling to others that they would like to start a drum circle. We may therefore be near the beginning of a performance, one that hasn't formally started yet and which will have its own rhythm of beginning, middle, and end. Moreover, the tradition of the drum circle in this particular park hit its height in the mid-2000s and is now long out of date: we are at a specific point in the growth and contraction of the art form. Not every audience member will experience every layer of this polyrhythm, but every audience member will experience more than one, and the result will make the drumming seem mournful ("look, a lonely holdout from the old drum circles! How sad!"), annoying ("can't even keep a beat, of course nobody

6 An extension of Levine (53), "The term *rhythm* moves easily back and forth between aesthetic and social rhythms" and her general interest in the collisions between forms. I extend the conversation in this way because I am interested in the rhythms of Bei Dao as they were encountered by their readers, rather than the rhythm as it is analyzed in the text. See also Kotz et al., in which they discuss the interaction of external rhythm with the rhythms of the receiving brain, explaining why "while rhythmic behavior is often based on quasi-periodic repetition of steady intervals, for musicologists and linguists alike a simple periodicity-based definition is incomplete" (896).

7 "Thick" in this context refers to Appiah (427), "translation that seeks with its annotations and accompanying glosses to locate the text in a rich cultural and linguistic context."

8 Meschonnic 145–46. He argues against distinctions between rhythm and meaning—the Bible, to him, means how it sings—and says that "to displace the usual relationship between rhythm and meaning implies a radical move from *totality* to *infinity*—which is nothing but the empirical condition shared both by history and by meaning. That move is also a move from linguistics, or semiotics, or pragmatics, as I know of them, to an historical anthropology of language." Translation, here, will be that "historical anthropology of language." For more on embodied translation as partial and even partisan translation, see also Admussen "Embodiment".

will join them"), or hopeful ("someone's got to start everything up again … "). These layers will interact with one another: the difference between the ideal beat and the performed beat, for example, will seem less important if the session is interpreted as a meaningless warm-up to a more organized circle that will start soon. These effects would multiply exponentially if there were several instruments and players: each ideal rhythm, each actual rhythm, and each interaction between each player and each other player. One can never reproduce a rhythm, its many influences, and polyrhythmic interactions; the intended beat would be easy enough to reproduce, but the rest is specific to its context in a place and time. What can be accomplished is to provide a readerly context, to elucidate important strands of the polyrhythm, and by doing so allow a reader to imagine what it is to hear a particular rhythm in a particular moment.[9] That particularity cannot be replaced by generalized abstraction: in the following, therefore, I will attempt to isolate and describe elements of a particular Chinese poem's polyrhythm, and to let readers see how they interacted with each other and with their readers.

In 1978, in the People's Republic of China, a group of young writers and painters posted their self-published literary magazine, *Today* (今天), across Beijing, including at the Democracy Wall at Xidan. Decades later, one can still hear the echoes of *Today*. Contemporary Chinese poetry is still centered around countercultural self-publication (van Crevel), it continues to be mostly free verse, like the poems in *Today*, and many poets self-identify by their temporal and ideological proximity to, or distance from, the "Obscure" or "Misty" (朦胧) poets that published in *Today*.[10] The standout work from the first issue of *Today* is editor Bei Dao 北岛's poem "The Answer" (回答), a free verse poem in seven stanzas.[11] Zhang Taozhou 张桃洲 (14) surveys the poem's reception, emphasizing its 1979 reprinting in China's most established official magazine and the role of later public discourse in establishing a politically "paradigmatic" role.[12] The heavily sociopolitical lens through which the poem has come to be

[9] Again, though, this process must be partial. The above example leaves out the interactions of the drum beat with the song of the birds or the swish of passing traffic. That omission is not evidence that these parts of the polyrhythm don't matter.

[10] The "third-generation" poets, for example, are the third generation of poets after the publication of *Today*. Zhang (13) talks about the early 2000s position taken by poets who want to "pass Bei Dao."

[11] A translation of the poem is available in Bei, *Rose* (7).

[12] He leaves this role unspoken, as his audience knows full well, but the specific politics that "The Answer" comes to represent is the Deng Xiaoping-led age of liberalization that followed after Mao Zedong's death and the denouncement of the Cultural Revolution Working Group, also known as the "Gang of Four."

seen, he insinuates, is artificial: "The influence produced by 'The Answer' has undergone a process, and its radiant power is primarily directed *backwards*" (14). Tang Xiaobing cites Zhang's ambivalent close reading of the sounds of the poem on the way to arguing that "The Answer" was a hinge moment in an "ocular turn" that saw the strident, orally performed political poetry of the Cultural Revolution supplanted by visual poetry that "moves away from a sonic impact or immediacy" (73). This interpretation, although it tracks Bei Dao's career and the general direction of Chinese free verse poetics quite well, does not answer Zhang's question about the backward-looking quality of "The Answer's" canonization. "The Answer," and the other poems in *Today*, mattered to people who read them in 1978, people who had no clear conception of the moment as a hinge. These were the people who read and spread the poem and provoked its later canonization. Understanding the poem's polyrhythms in their moment is crucial to understanding the poem's interaction with its moment, a perspective that Zhang points out is consistently overlooked.

Rhythm is physical; it is a material pattern. For most Chinese readers in 1978, the poem's rhythms were created by the print journal *Today*, original copies of which have been digitized by the library at Leiden University.[13] The cover has the journal's Chinese title, 今天, but its English translation is not *Today*, as it would be starting in issue 3, but *The Moment*—a singular and particular slice of time that does not move constantly into the future (as today does). The text was produced with mimeograph machines whose characters were set by being etched into wax. From a contemporary perspective, they seem homespun and crude, but from the perspective of nongovernmental actors in 1978, this was an almost shocking seizure of the normally state-controlled means of literary production. When *Today* editor Mang Ke 芒克 first saw Bei Dao's initial, experimental mimeograph publication of his own collection of poems in the mid-1970s, the fact that they were printed seemed almost frighteningly transgressive to him (Mang 8). He himself had hand-copied his poems for *samizdat* circulation among his peers, and then later burned them in advance of a wave of arrests that took place in 1973. The amount of paper on which *Today* was printed was illegal to purchase without special permission, and so members

[13] Leiden's Unofficial Poetry Journals from China (UNPO) project has scanned two different copies of the first issue of *Today*, with minor differences (and these differences are themselves evidence of the hand-made, individually manufactured process of making early *Today*): I use the one I find slightly clearer, accessible at http://hdl.handle.net/1887.1/item:3177787. "The Answer" is on pages 28 and 29.

of the community visited stationary stores repeatedly, buying smaller amounts until they had enough to run the journal (E Fuming). Printing was done secretly at a house near Beijing; editors and their friends worked around the clock, and tried to enter and leave as little as possible (Mang 95–7). Editor Mang Ke, in his memoirs, almost shakes his finger at contemporary readers who think of *Today* as a purely conceptual act: "可别小瞧油印这活儿," he writes, "you really can't underestimate the work of mimeographing" (97).

The hand-carved characters, cheap ink, and schoolhouse paper of early *Today* were most important because of the way they signified to readers. In their first incarnation, posted across Beijing alongside other mimeographed journals of the period like *Beijing Spring* (北京之春), they were interpretable as a form of Cultural Revolution-style big character posters, grassroots communications that the communist headquarters had once encouraged as a method of attacking entrenched, antirevolutionary elements inside the CCP. *Today* contributor Wanzhi (万之, pen name of Chen Maiping 陈迈平) describes it like this:

> The political environment in Beijing at that time was similar to the early days of the Cultural Revolution. There were a lot of big-character posters, small-character posters, headless posters, and mimeographed publications on the Xidan Wall. In universities, students were also eager to try. Students in the Chinese Department even set up their own poetry clubs and I was very busy with the mimeographing of poetry. (n.p.)

Rather than a hinge moment bending toward a novel future, to readers at the time early *Today* felt more like a formal return to early Cultural Revolution media tactics; those who would become *Today*'s first subscribers stood, in the Beijing cold, reading texts publicly posted on a wall just as they had at the beginning of the Cultural Revolution (Cheng). Rather than defeating the party and striking out for new forms of government, the form was a way to engage in factional intraparty struggle, and indeed, parts of the communist apparatus would eventually print formal versions of *Beijing Spring* and, as Zhang Taozhou points out, reprint poems from *Today* in high-circulation official literary magazines.[14]

[14] Much of the data from memoirs indicate that poets were, in addition to taking influence from foreign poetry and the work of early 1970s underground poets such as Shizhi 食指, inspired by or participants in the 1976 Tiananmen demonstrations, in which demonstrators left poetry and prose in memory of perceived centrist Zhou Enlai, and his political supporters. Mang Ke (61) seems a bit mystified that people today no longer talk about it, as it was an exciting ("够热闹") moment.

The mimeograph of *Today*, which was made available by mail subscription, indicated that the young liberals and avant-gardists of the cities were not being arrested for sedition, and were thriving. One recipient of the magazine from Jilin province called it "comforting" that such a thing was possible (Xu n.p.).[15] The comparatively high level of technology with which the *Today* group reproduced their art, and their ability to take mail subscriptions, announced the strength and semi-legality of their faction, a representation of strength and boldness that would have been recognizable to those who lived through the age of the Red Guards. For people at the Democracy Wall, citing this early-1960s moment was a practice of historical rhyme that is common to Chinese aesthetic and political speech; demonstrators at Tiananmen in 1976 would make intentional rhymes with the May Fourth movement of 1919, and protestors at Tianamen in 1989 would rhyme with 1976. The visual prosody of the poems—the way they were made on the page and the way the page was displayed to readers both by public posting and by subscription—accomplished that historical echo, and the rhythm of reading created a historical rhythm that mattered deeply to the reception of "The Answer."

As Wanzhi points out above, however, in the late 1970s lots of young Beijingers and others were trying to mimeograph poetry, and not many of them had the impact that *Today* would have. The magazine is often described as a shocking innovation, but part of the shock came from the journal's revelation of years of underground poetics. *Today*, especially its first issue, is a collection of poems that had been written, shared, exchanged, and revised for years before they appeared in print. Chen Chang 陈昶 gives a comprehensive version history of "The Answer," arguing that it split off from an earlier poem "I'll Tell You, World" (告诉你吧, 世界) in 1973, and reached its basically mature form in 1976 (Chen). Mang Ke reports that 1973, especially before the wave of raids and arrests, was the most energetic period of underground composition in Beijing (41), with poets writing new work, reading it to friends, and bringing it to poetry salons, iterating it, and seeing which of their poems would be most successful with their peers. Success could mean praise at a clandestine reading or positive

15 This feeling was perhaps reinforced by the way that 油印 *youyin* or "oil-printed" mimeographs were both machine-produced and hand-made: each page was individually loaded and printed, and each one has imperfections, giving concrete traces of human production. These indicate not simply that liberal or antirevolutionary people have access to mimeographs, but that they have taught themselves, and are still teaching themselves, how to operate them. A much more precise printing might not have felt so open to imitation.

word of mouth—Mang Ke remembers hearing about specific poems long before he had the chance to hear them performed (22)—ultimately, it often meant the circulation of hand-copied versions of a poet's work. The writer Ah Cheng 阿城 recalls being sent down to Inner Mongolia during the Cultural Revolution, and finding a new friend who had Shizhi 食指 poems he'd never seen before (Ah Cheng). The same kind of young 1970s thinker who wrote for *Today* was also often engaged in fiction writing, visual art, and musical appreciation, but it was poetry that was best adapted for travel through the culture of underground salons, and the quality of the poetry in *Today* could not but be affected by the years of work and discussion that had surrounded its composition. So the publication of "The Answer," alongside the other poems of *Today* in mimeograph across Beijing, was not, exactly, an innovation: it was the publicization of an ongoing, established experimental poetry scene that had been engaging in a specific form of verbal and visual poetics for many years. The heat and energy with which "The Answer" was received comes in part from the way the written text insists on an oral performance, the way it allows readers to imagine themselves being read to, announces a real and preexisting community of writers, and initiates the reader into membership. Tang Xiaobing writes,

> The sonority at the heart of the poem, which has received much critical acclaim, is therefore deeper than is made possible through a defiant tone, rhythmic lines, or unequivocal pronouncements. It stems from the mode of reading that, prescribed within the poem, stipulates we activate our aural imagination and hear an urgent but firm voice rising from the text …. (68)

This is a result of the rhythms of poetic life in the Beijing underground: one writes a poem, performs it, and attempts in the process to get others to copy and perform it. The pieces they selected for their journal were the gems of a poetry culture aimed substantially at provoking autonomous circulation. It is no wonder that the aftermath of the publication of *Today* saw a generation of its readers turn into underground poets, editors, and collectors. One was not intended to just buy and consume *Today* but to recite it, circulate it, imitate it, submit art to it, and even to join up in its production. After the first issue was published, the editorial board held public readings, even though they knew they would be identified, photographed, and followed (Ah Ming). Readers left notes at sites where *Today* was posted, trying to get in touch with its community (Mang 108). Rhythm is not just comprised of the call but also the response: in this case, reader response was provoked in part by a written text that insisted on oral declamation, an organization of words on the page that insisted that they be read. In Jilin, the writer Xu Jingya who found the journal "comforting" lent it out, recopied it, and most of all hosted readings where it was publicly performed (n.p.).

With these contextual elements of the polyrhythm laid down around the poem, we are more prepared to talk about "The Answer's" textual rhythm. This account will focus on the two most rhythmically crucial parts of the poem, the opening, in which it announces its form and shape, and the poem's celebrated fourth stanza. The poem opens like this:

> 卑鄙是卑鄙者的通行证,
> 高尚是高尚者的墓志铭,
> 看吧,在镀金的天空中,
> 飘满了死者弯曲的倒影。(*Jintian* 28)[16]

> Debasement is the password of the base,
> Nobility the epitaph of the noble.
> See how the gilded sky is covered
> With the drifting twisted shadows of the dead. (Bei, *Rose* 7)

This is a tightly rhymed quatrain with multiple rhythmic features from classical poetry: the lines end in zhèng, míng, zhōng, and yǐng, providing an xAxA rhyme and two more near-rhymes from the consistent use of -ng finals. The quatrain has ten syllables per line—with one important exception—and sounds audibly regular. Syllable counts in classical poetry are usually four, five, or seven syllables per line, but the more verbose, hypotactic sentences of modern Chinese fit better into an eight- or ten-syllable line (Admussen, "Music" 18). What's both audible and visible about the poem's rhythm, though, is that the opening lines are syntactically parallel, with line one and line two having the following identical structure: two-character adjective, copula, three-character nominalization of the two-character adjective, particle "de," three-character noun. This is a truly ancient hallmark of patterned Chinese writing, and even those who rejected the great Tang dynasty poets as side effects of feudalism would have recognized these structures from the classical-style poetry of Mao Zedong. To almost the

[16] Some later versions of the poem insert the relative pronoun 那 after 在 in line 3 of the stanza, changing both the visual and the oral prosody; we are here interested in the poem's first open publication as collected at Leiden.

maximum extent possible, in its first few lines "The Answer" announces itself as a deeply patterned piece of Chinese poetry, which is why it comes as a real shock to hear line three's highly vernacular direct reader's address kàn ba, and the dramatic, highly audible comma caesura after it. The English translation doesn't emphasize the word, but it would be fair to translate it as "Look!" or "Take a look," which would give a better sense of the poet shifting from a declarative, philosophical register to a much more personal type of voiced presence. The presence of the comma stands in for a Chinese character in the visual prosody of the meter, keeping the lines even and neat, but the length of a caesura, especially this one, does not equal a single syllable's *mora*, which in Chinese is much closer to standard than in accentual languages like English. The result is that the line looks visually metrical—it's the same length as the others—but it sounds voiced and vernacular when read aloud. The drama of this moment, in which a static, structured, recognizably poetic couplet turns to a verbal and vernacular act of leaning forward toward the audience, would not have been experienced as a hinge toward a future of visual poetics. It instead insists on the sonority and social role of printed poetry, the expectation that seeing a poem will make one want to say it. This is accomplished through a shift in audible rhythms that is covered by a highly structured visual form. The shift is also in mode, from a pronouncement to an oration, from the philosophical depth of traditional Chinese poetry to the Red Guard (or in this case, demobilized and reformed Red Guard) standing in front of a big-character poster, urging others to join a cause.

The poem's most oft-quoted (we might today say most viral) stanza is stanza four, but to read its rhythm we should start in stanza three:

> 我来到这个世界上,
> 只带着纸、绳索和身影,
> 为了在审判之前,
> 宣读那些被判决了的声音。
>
> 告诉你吧, 世界,
> 我——不——相——信!
> 纵使你脚下有一千名挑战者,
> 那就把我算作第一千零一名。 (*Jintian* 28)[17]

[17] The 1978 version of the poem in *Today* at the Leiden archive (in both the variants they collect) seems to be missing the bamboo radical at the top of the character 算; it appears in Bei Dao's holograph manuscripts (Chen) and so I've changed it here.

I came into this world
Bringing only paper, rope, a shadow,
To proclaim before the judgment
The voice that has been judged:

Let me tell you, world,
I—do—not—believe!
If a thousand challengers lie beneath your feet,
Count me as number thousand and one. (Bei, *Rose* 7)

Stanza three opens with abstract symbols that would be at home in the translated European poetry that was fashionable in the underground salons of 1970s Beijing,[18] and then, as we've seen before, transitions to an insistent oratorical mode. Like the opening stanza, it uses the imperative to give a sense of a physical listener, and uses punctuation to slow down the recitation, to make it sound deliberate and unequivocal. The fourth stanza's long dashes are fascinating pieces of visual and oral technology: horizontal dashes, including hyphens and em-dashes, are nearly identical to the character 一 *yi*, the number one, and so the use of both long and short dashes is exceptionally rare in most Chinese works, and is most commonly seen in translations of foreign literature where the reason for the awkwardness is clear.[19] However, in holograph manuscripts, as well as wax-etched manuscripts like that of *Today*, the dashes can be written off-center or tilted to indicate that they are not the number one. The first issue of *Today* puts the dashes about halfway between the center of the line and the place where one might typeset an underscore. The result is odd, and feels simultaneously handwritten, translingual/transcultural, and overwhelmingly oral. In part because the syntax and the poem's self-identification as proclamation provide a guide, though, nearly every reader would know what to do with line two of stanza two, and even readers unused to translated literature in the 1970s would have performed an almost full stop between every word. In this pair of stanzas, then, we have an experimental, partially imported visual form pushing the

[18] Bei ("Translation Style") describes how these texts circulated, what they meant to underground writers as an alternative to Maoist linguistic formulations, and which writers were most esteemed in his circles.

[19] Note please that dashes are much more common in Chinese now; this is, to my mind, partially a result of the circulation of "The Answer." The typographical problem has been solved by making the dashes two uninterrupted character-widths long, preventing them from being misinterpreted.

reader toward a rhythm from socialist oration. And underlying this visual/verbal interaction, we have the rhythms of aesthetic life in the coterie salon: I perform a poem for you, we make a text copy of the poem together, and that text travels to another salon where someone reads it aloud again. What is important about the poem, the reason we read and attend to it today, is the way that it successfully reproduced the rhythms of salon circulation on a much larger scale.

Bei Dao has all but repudiated the poetics of "The Answer." During his productive and storied later career in exile, he wrote

> Now when people bring up "The Answer" I often feel ashamed. I have a basically negative attitude towards that kind of poetry. In lots of ways, it is an echo of official language. At that time our writing had a very intimate relationship with revolutionary poetry, so much of it was high-pitched, used big words, and held violent tendencies. We came from that era, we can't avoid being influenced by it .... (Bei "Chinese")

These are reasonable statements, opinions widely held by artists who came of age in the 1970s, and evidence of Tang Xiaobing's argument about a hinge from oral to visual prosody. "The Answer" is not adapted to the struggle against neoliberalism or the cultural depredations of globalization, it is not the shape of contemporary Chinese verse in 2003, when Bei Dao gives the above interview; it is perhaps not the right poem *for those moments*. But implicit in his shame about the poem is his acceptance that poetry at that time had to be written inside the strictures and limitations of that time. The poetry of the 1970s had to confront and transform the rhythmic expectations and prosodic knowledge of its readers, it had to appear physically and aurally according to the medial technologies of the time. And in its moment, "The Answer" was a masterpiece of polyrhythm, announcing itself as a part of a very old and morally authoritative poetic tradition, leaning forward to address its reader in a strident, intimate voice, defamiliarizing the line and even the Chinese character with novel caesura, and appearing just exactly at the historical moment in which these rhythms first became possible, in fact announcing their possibility to a readership habituated to a quite inflexible literary system.

Appreciating "The Answer" today, through a thick translation of the layering of its social, political, visual, and verbal rhythms is *not* valuable as an objective assessment of the poem's value in 1978. To explicate rhythm in this and every case is also to create an additional polyrhythm, laid over the poem, that provokes resonances in a contemporary audience. "The Answer," especially when tested against the more cleanly modernist and visual poetry written after the Chinese turn toward the ocular, shows that the vitality of hybrid identities, forms, and politics comes in part from the complexities of the polyrhythm. Bei Dao, having grown up in a life-or-death negotiation with an authoritarian state and then traumatically transculturated to exile outside it, has attempted to clean socialist poetics from his work. But "The Answer" uses the socialist poetics of organization, and strident affect, to move and organize readers in a way that later poems simply did not. Because of its polyrhythms, because the manner and moment of its publication meant that the communities it organized were independent of the state literary apparatus, the poem's effect was emphatically not to encourage obedience and piety to the Chinese Communist Party: "The Answer" was chanted at Tiananmen in 1989 and the aesthetic descendants of the Obscure poets remain a source of freethinking and independence in China today. This continued cohesion is partially enabled by the willingness of the poets of the 1970s to mix and add rhythms: visual, verbal, foreign, classical, socialist, and experimental.[20] The goal, in that context, was not ideological purity but the response of the salon, the coterie, and eventually the city and nation. "The *polis*," Caroline Levine writes, "cannot do without form" (xiii). And a superabundance of rhythms, an instrumental rather than ideological approach to rhythm, makes a richer and more flexible form in which the polis can reside.

This observation brings me to an unintended conclusion, a full-throated defense of presentism in the teaching and study of rhythm. Because rhythms are shapes in time, and because those shapes are inextricable from the moments in which they appear, to reproduce a historical rhythm for analysis is necessarily to translate it into the present, to reperform it, and to make something new. Polyrhythm's affordance for endless layering, however, decreases the risk of erasure that presentism can threaten: layers of reperformance and transformation are additive rather than competitive, and one does not need to alter or revise a text's archive to perform it in a contemporary context superimposed over other contemporary rhythms. What presentism justifies and encourages, rather, is attention to the arrangement and selection of texts, the dissonances and echoes their rhythms might provide, and especially the *uses* to which contemporary readers and students might put the rhythms reperformed. I choose "The Answer" to bring to this moment because I am fascinated by, and strongly feel the need

[20] In American poetics, there may be a similar vitality created by contemporary poets such as Danez Smith who are participants in spoken word scenes, but who also publish in print.

for, the poem written for friends, the speech of the small, unofficial salon.[21] As a counterpoint to the viral text, which slides frictionlessly between contexts, "The Answer" teaches the poetics enabled by reading and inhabiting a specific context. And where Bei Dao's perception of his community in the 1970s was perhaps set against the "big we" of a community that claimed too much authority over individual subjectivity, today in my classrooms "The Answer" feels like a hinge backward toward the messy, social intersubjectivity of local organization. Pointing this out, and caring about it, does not change the rhythms of the poem: it simply attends to their rhythmic arrangement in the temporality that surrounds them, which in this essay is the present. This additive superimposition, and its endless, lossless iterability, justifies the use of translation as a method for the investigation of rhythm: like a poem in a foreign language, a historical rhythm can be translated and retranslated, remade and remade, refreshed and opened to multiple social use by its shifting relationship with the ever-moving flow of time.

Works Cited

Admussen, Nick. "Embodiment in the Translation of Chinese Poetry." *Chinese Poetry and Translation: Rights and Wrongs*. Edited by Maghiel van Crevel and Lucas Klein. Amsterdam UP, 2019, pp. 113–34.

Admussen, Nick. "A Music for Baihua: Lu Xun, *Wild Grass*, and 'A Good Story.'" *Chinese Literature: Essays, Articles, Reviews*, vol. 31, 2009, pp. 1–22.

Ah Cheng 阿城. 《昨天今天或今天昨天》. 《今天》40 年, 2018. www.jintian.net/today/?action-viewnews-itemid-70414.

Ah Ming 阿鸣.《记《今天》编辑部的一次诗歌朗诵会》. 《今天》40年, www.jintian.net/today/?action-viewnews-itemid-70468.

Appiah, Kwame Anthony. "Thick Translation." *Callaloo*, vol. 16, no. 4, 1993, pp. 808–19.

Bei Dao. *The Rose of Time*, edited by Eliot Weinberger. New Directions, 2010.

Bei Dao. "Translation Style: A Quiet Revolution." Translated by Wei Deng. *Inside Out: Modernism and Postmodernism in Chinese Literary Culture*. Edited by Wendy Larson and Anne Wedell-Wedellsborg. Aarhus UP, 1993.

Bei Dao 北岛, 2003. 《中文是我唯一的行李———北岛访谈》 (Chinese is my only baggage—an interview with Bei Dao), 《书城》 no. 2.

[21] This is a fit moment to thank the members of the Cornell graduate program in China Studies: the underlying idea for this essay was assessed and validated by a seminar where Lin Chengxiang, Kevin Dong, Hong Xiao, and Bruce Li read and discussed Tang Xiaobing's essay. My intellectual production as a scholar is dependent on this and other small-scale communities of interlocutors and friends.

Chen Chang 陈昶. "北岛早期诗歌源流考——以《告诉你吧, 世界》与《回答》的版本问题为核心." (A study of the origin of Bei Dao's early poetry: with the versioning issues of "Listen to Me, World" and "The Answer" at its core.) *Yangtze Jiang Literary Review*, 扬子江文学评论, vol. 3, pp. 54–60.

Cheng Qifeng 程奇逢. 忆《今天》. 《今天》 40 年 2018. www.jintian.net/today/?action-viewnews-itemid-70438.

E Fuming 鄂复明. 《早期《今天》事务摭忆》. 《今天》 40 年, 2018. www.jintian.net/today/?action-viewnews-itemid-70436.

Handel, Stephen, and Gregory R. Lawson. "The Contextual Nature of Rhythmic Interpretation." *Perception & Psychophysics*, vol. 34, no. 2, 1983, pp. 103–20.

Jintian 《今天》(1978). Edited by Bei Dao 北岛 et al. vol. 1, no. 1, Beijing. http://hdl.handle.net/1887.1/item:3177787.

Kotz, Sonja A., Andrea Ravignani, and William T. Fitch. "The Evolution of Rhythm Processing." *Trends in Cognitive Sciences*, vol. 22, no. 10, 2018, pp. 896–910.

Levine, Caroline. *Forms: Whole, Rhythm, Hierarchy, Network*. Princeton UP, 2016.

London, Justin. *Hearing in Time: Psychological Aspects of Musical Meter*. Oxford UP, 2004.

Mang Ke 芒克。《往事与《今天》》(The past and Today). Xinbei: INK Literary Monthly Publishing Co., 2018.

Meschonnic, Henri. "Translating Biblical Rhythm." *Modern Language Studies*, vol. 15, no. 4, 1985, pp. 143–56.

Saussy, Haun. *The Ethnography of Rhythm*. Fordham UP, 2016.

Tang Xiaobing. "The Ocular Turn, Misty Poetry, and a Postrevolutionary Imagination: Rereading 'The Answer' by Bei Dao." *PRISM: Theory and Modern Chinese Literature*, vol. 16, no. 1, March 2019, pp. 62–84.

van Crevel, Maghiel. "Transgression as Rule: Freebooters in Chinese Poetry." *At the Shores of the Sky*. Edited by Paul Kroll and Jonathan Silk. Brill, 2020.

Wanzhi 万之 (2018). 也忆老《今天》.《今天》40年. www.jintian.net/today/?action-viewnews-itemid-70440.

Xu Jingya 徐敬亚. 《中国第一根火柴. 《今天》 40 年, 2018. www.jintian.net/today/?action-viewnews-itemid-70466.

Yang Zhiyi. "Classical Poetry in Modern Politics: Liu Yazi's PR Campaign for Mao Zedong." *Asian and African Studies*, vol. 22, no. 2, 2013, pp. 208–26.

Zhang Taozhou 张桃洲. 《重读北岛的《回答》》. 廊坊师范学院学报 (社) no. 3, September 2016, pp. 13–16.

11

Rhythms in Painting (1910–30)

Georges Roque

To Mark Lussier, remembering those days—forty years ago, yes forty—when we intrepidly held our First International Congress on Rhythm with the two of us as unique speakers under the majestic ruins of Tulum and soothed by the cadence of the waves smoothly hitting the turquoise water of the beach …

Georges

Are there rhythms in painting? It obviously depends on the way rhythm is defined, which necessarily raises the huge issue of what rhythm is, as there is no consensus among scholars about its definition. So maybe we could ask instead: Why are there so few papers on rhythm in painting within the growing field of rhythmic studies? Here the answer is easier, as some consider that there is no "proper" rhythm in painting. To give just one example: "There are only three fields in which the concept of rhythm finds its ontological and empirical reality: biological organisms (bodily movements and physiological functions), language and music. This is the reason why the other uses of the word rhythm are mere metonymies" (Fantini 217). Such a view is widely shared (see, for example, Gaiger). If it were true, rhythm in fine arts like painting would be impossible unless rhetorically. This might explain the lack of studies on this topic, with few exceptions (in particular in the field of semiotics; Enríquez and Lopes; see also Tosaki, "Rhythm as Schema," who proposed an interpretation of rhythm as scheme, which can be applied to visual static arts). Fortunately, painters are less timorous than art historians.

In this chapter, I will focus on the beginnings of abstract art[1] insofar as rhythm is indeed much discussed among the pioneers of nonfigurative art (Visnadi).

[1] For an overview, see Roque.

One might ask why. The best answer was probably given by Kandinsky in *On the Spiritual in Art*, from comparisons with music, as he claims that "the richest lessons are to be learned from music" (154). Music serves as his model, insofar as, according to him, it is the art that uses its means "not to represent natural appearances, but to express the inner life of the artist and create a unique life of musical tones" (154). Indeed, an artist who no longer seeks to imitate nature but wants to express his inner world "sees with envy how naturally and easily such goals can be attained in music, the least material of the arts today" (154).

After this very clear statement, which also applies to many other artists, and summarizes a common concern at the time, as soon as the goal of imitating nature moves away—which explains the importance of musical titles in the paintings of many pioneers of abstract art—Kandinsky adds: "Hence the current search for rhythm in painting, for mathematical, abstract constructions, the value placed today upon the repetition of the color-tones, the way colors are set in motion, etc." (154). These quotes help to understand better the importance of rhythm in abstract painting, beyond Kandinsky.

When I started working on this topic, I first wondered whether this interest in rhythm was manifested in the titles of the works. From this point of view, empirical research is rather disappointing: we find in fact relatively few titles referring explicitly to rhythm, although titles alluding to music abound, for example in van Doesburg's paintings. The main exception concerns the Delaunays, in Robert's works like the series *Rythme sans fin* and then in Sonia's.

Now, the relationships of abstract painters to the idea of rhythm are numerous and complex, and the material is very abundant. As the titles of the works can hardly serve as a guide, we will turn here above all to the texts of the pioneers commenting on the idea of rhythm. This eliminates a good number of artists who have not conceptualized the idea of rhythm. This is particularly the case for Robert Delaunay. He most often uses the term "rhythm," but in a loose meaning that is not very useful. I will therefore quickly review the writings of some artists from the beginning of the twentieth century (Survage, Kupka, Mondrian, Kandinsky, and Klee). My aim is above all descriptive, as these writings have never, as far as I know, been compared.

Survage

Like many others, Delaunay considers that rhythm is linked to movement. Survage is a good example of this attempt to link rhythm and movement. However, for

Delaunay and other painters, the issue is to give a feeling of movement from still images. Survage had a different ambition: we know that he planned to make a film from several hundred watercolors, to give movement to color, in the sense of a succession of images taking place over time. It seems that Gaumont was interested in supporting the film, but the First World War interrupted the project. A certain number of drawings on paper remain, dating from 1912 to 1913. Survage called his project "Colorful Rhythms" and published a paper under this title in the *Soirées de Paris* (in July–August 1914). In this text, Survage starts from the analogy between color and music, the two being formed by vibrations. He was to clarify these ideas in a memoir submitted to the Academy of Sciences on June 29, 1914, entitled "Color, Movement, Rhythm," in which he explained that painting, having freed itself from the representation of nature, must still free itself from immobility, to become, he said, "a means as flexible and rich to express our emotions, as is the case with music. [. . .] I animate my painting, I give it movement, I introduce rhythm into the real action of my abstract painting, hatched from my inner life, my instrument will be the cinematographic film" (Survage 25). We know that Apollinaire and Blaise Cendrars, very enthusiastic, wrote texts praising Survage's project.

His is an interesting case for several reasons. The analogy with music is obvious, because for Survage "music is a mode of succession *in time*, of various sound vibrations" (21, original emphasis). It is this presence of color movement taking place in time that he wanted to render thanks to cinema. Vibrations form another common point between music and color, as well as the issue of states of mind, as for him "a musical work is a kind of subtle language in which the author expresses his state of mind" (21). According to him, there is a difference between the effect produced by an immobile and a mobile abstract form. Immobile "it produces only an extremely confused sensation, it is only a mere graphic notation. It is only by setting itself in motion, by transforming itself and by meeting other forms, that it becomes capable of evoking a feeling" (22). We therefore find here the idea of associating rhythm and movement to produce feelings in the spectator.

Kupka

Not everyone shares this radical conception implying a real movement in painting to obtain a rhythm. Kupka, for his part, also links rhythm and movement, but thinks of obtaining it directly on a canvas. For him, as he clearly says, it is a matter of: "giving the impression of movement by using means which are themselves immobile!" (Kupka 198). Indeed, Kupka thought that actual movement could be obtained in painting thanks to the physiological phenomenon of retinal persistence (responsible, by the way, for our perception of movies). In his book, probably written between 1910 and 1913, he explained in a chapter devoted to rhythm:

> Lines and the value of area are the main features implemented by the artist. If he wants to mark a rhythm, he must take into account the length of the impressions produced by these two factors. [...] In order for a work to be rhythmed, its lesser parts must contribute to it, be controlled, be requested and match each other. Rhythm resides in periodical return of analogies, in the symmetry of lines and limits. (Kupka 196)

Even if Kupka insists on retinal persistence as a way of introducing rhythm into painting, at the same time he proposes pure plastic means of suggesting rhythm (repetition of analogies, symmetry). From what precedes, one could think that Kupka had a purely formal conception of rhythm in painting. But it is not so. Insofar as Kupka conceived of color as deeply related to mental feelings and emotions, and these in turn being physiological states, he then considered rhythm as "organic rhythm" (162). About the stroke, he wrote that painters had to take as an example the Japanese practice: "the breathing rhythm must correspond as far as possible to the rhythm of the very act of drawing in such a way as never having a complete stop" (171). Trained within Viennese symbolism, Kupka had an enlarged vision of rhythm, related to sensations and states of mind, to organic rhythm (in the above quote, it deals with breathing, but other passages are about blood composition), and finally to cosmic rhythm:

> All probably lies in the cadence danse of punctual sites, producing impressions we perceive as a discontinuous succession. If we sound its principle, we will find that it is similar to the life of an elementary cell: its dilatation and contraction movements correspond to the cosmic rhythm of reproduction and return. (199)

Kandinsky

For what concerns Kandinsky (Figure 11.1), rhythm is far from being central in his writings. It doesn't appear in *Reminiscences*, not even in the important

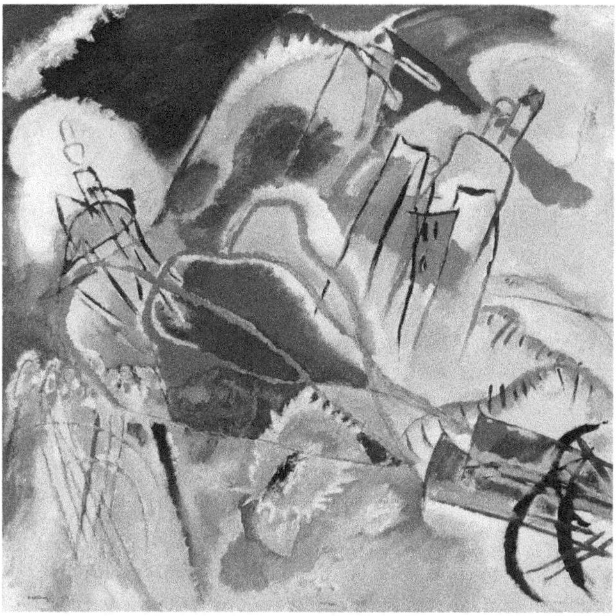

Figure 11.1 Vassili Kandinsky, "Improvisation No. 30 (Cannons)," 1913, oil on canvas, 111 × 111.3 cm. Art Institute of Chicago: CC0 public domain.

comments he gave us on some of his paintings: when analyzing them he uses the concepts of harmony, equilibrium, contrasts, masses, but not rhythms. Yet the most interesting for us are the notes for his teaching at the Bauhaus. After completing *Point and Line to Plane*, he more directly deals with the issue of composition narrowly bound to that of rhythm. His notes (unfortunately there is no written book available)[2] contain many references to rhythm, as seen in different ways. One is rather classical: "Rhythm is, ultimately, synonym of proportion; the essential way of expression of all the arts. Rhythm defines proportion, and even an architecture can be in D major or in E-flat" (Kandinsky 139–40). According to this conception, also applied to architecture, "right proportions make men happy" (140).

Another, more important, idea is based on tension: "rhythm = absolute relationship between tensions [...] same thing for all the arts: painting, music, danse, sculpture, poetry, architecture" (Kandinsky 135). Rhythm,

[2] A former student of Kandinsky at the Bauhaus, Jean Leppien, had access to these handwritten notes and translated them into French (*Cours*). As far as I know, they haven't been published in German, and are not translated into English. In what follows, I will translate them from French into English.

tension, and composition are then narrowly intertwined. "Main principle of composition? Answer: rhythm. Rhythmic construction through lines, shapes and colors, relationship to the plane" (169–70). Elsewhere, he wonders: "what is rhythm?" and answers: "the relationships between tensions. The law of rhythm? Tension, relaxation = contrast. All is relative, hence the possibilities are endless" (168). But what exactly are tensions? In other notes, the painter explains:

> The tensions inherent in elements are static forces waiting for encounters making it possible for them to become active = dynamic. Tensions between two lines prevail over small initial tensions and produce a beat. As soon as we perceive this beat, the initial tensions become secondary. The simple and primary tensions, from one element to the next, create *rhythm*. It is obvious that rhythm define the work, that a work without rhythm is inconceivable. But the rhythm is not a more or less simple repetition but the composition of tensions between elements = *composition of relations*. (226, original emphasis)

Kandinsky's conception of rhythm is far from being limited to tensions within composition. It goes beyond. "What is the absolute law of composition? A rhythmic, logic and effective construction = laws of nature and until cosmos" (Kandinsky 168). We find a concept parallel to that of Kupka, as rhythm is not limited to composition. It is precisely about composition that Kandinsky wrote:

> the first and primordial principle is rhythm: pulse, breathing, blood circulation. Is earth breathing? Is coherence of universe purely mechanical? Breathing of man, animals and plants coincide with the breathing of cosmos. Inaudible but precise "music". It is therefore natural that every human creation—including art—pertains to the same cosmic pulsation. So every real artist depends on nature, even if he is "abstract." (235)

He then resumed the idea that rhythm in art is a relation between elements and an alternation of tension and relaxation.

Mondrian

Another pioneer of abstract art who developed ideas about rhythm is Mondrian. With him we are nevertheless in a quite different worldview. The main difference is that Kandinsky's cosmological conception of rhythm doesn't make any difference between rhythm in nature and in art. They both belong to the same rhythmic vision of tension and alternating of tension and relaxation. Yet Mondrian was very worried by the construction of an art completely separated from nature. Hence the opposition he stresses between particular and universal. Art must be detached from the particularities of the natural world to embrace universal relations. For the same reason, he decided to reduce, in his paintings, the relationships between lines to the right angle and those of colors to the three subtractive primaries (Figure 11.2), as is well known. He considered that works of art made from these relationships would be more universal, as they don't borrow anymore from nature. This is why, for instance, he abolished green from his palette, because it is pervasive in nature, and then too particular.

Figure 11.2 Piet Mondrian, "Study for a Composition," 1940–1941, collage of cut and pasted papers, 33 × 2 cm. Art Institute of Chicago: CC0 public domain.

His dissociation between nature and art has a crucial consequence on the way he conceived of rhythm. Insofar as rhythm is present everywhere, Mondrian was led to distinguish and dissociate two conceptions of rhythm: that of nature and that of art. Consequently, as he wishes to produce artistic forms quite separated from nature, he was led to praise rhythm in art and, accordingly, to reject rhythm in nature.

Up to now, this chapter has been merely focused on the artists writings. In the case of Mondrian, however, an important essay devoted to his understanding of rhythm, Tosaki's *Mondrian's Philosophy*, must be discussed, even if my main purpose is not to comment on the possible sources of Mondrian but rather to give a summary of his ideas to better compare them with the other artists under scrutiny. It must be said that I disagree with Tosaki on several points. First, the importance he gives to theosophy as a source for Mondrian's ideas on rhythm (Tosaki, *Mondrian's Philosophy* 4, 13–21) seems to me exaggerated. Furthermore, he considers that there are "two sources of philosophical insight for Mondrian: mysticism and empiricism" (6). Would that mean that Mondrian would have read philosophical essays on empiricism? In fact Tosaki explains that "a major source in itself was his experience during his career as a painter" (8). That's true, but can his own practice as a painter be considered as a "source," at the same level as theosophy? As happens (too) often, it is easier to look for written sources than understanding the empirical practice of a painter.

More importantly, Tosaki considers that there is an evolution in Mondrian's ideas of rhythm, from a negative to a positive one (*Mondrian's Philosophy* 47, 58), from a static to a dynamic one (46, 63). However, for me, there is no contradiction and no ambivalence in Mondrian's understanding of rhythm. As explained above, the difference between a negative and a positive idea of rhythm arises from the important distinction he stresses between nature and art. He already insisted on it in one of the first papers of the series he published in the magazine *De Stijl*, the fundamental one devoted to plastic means and composition:

> In the new plastic the law of *proportion* leads the artist properly to realize the relationships of size and color on the picture plane: *purely and simply through plastic means* and not by any pictorial artifice. Rhythm becomes determinate: naturalistic rhythm is abolished. (Mondrian, "L'art nouveau" 40, original emphasis)

To the benefit of which other concept must rhythm in nature be abolished? Mondrian makes it explicit in the next sentence:

> Rhythm interiorized (through continuous abolition of oppositions of *positions and size*) has nothing of the *repetition* that characterizes the particular; it is no longer a *sequence* but a *plastic unity*. Thus it renders more strongly the cosmic rhythm that flows through all things. ("L'art nouveau" 40, original emphasis)

And he adds in a footnote: "The ancient Chinese conceived of rhythm as the life-fluid." As in Kupka, we find here the influence of the oriental understanding of rhythm. It turns out, then, that, to repeat it, there are two different rhythms, one in nature and the other one in the arts, and more precisely in what he will call neoplasticism. From this point of view, he never abandoned a negative understanding of rhythm (repetition pervasive in nature) in favor of a positive one (a plastic unity), so that it is useless to try to find out when exactly he would have shifted from the first one to the second one. He rather conceived of rhythm as double and different when seen in nature or in his neoplastic paintings.

Surprisingly, it is not in his writings about relationships between music and painting, but rather in his manuscript from the beginning of the 1930s, *The New Art—The New Life*, that Mondrian developed his notion of two rhythms, as linked to the complex issue that stands at the heart of his thought: the relationship between plastic expression centered on the representation of nature (which remains particular), and purely plastic expression (which reaches the universal). Mondrian's insistence on rhythm is due to the fact that it lays on both sides, but it is not the same one. Rhythm is then a crucial element already present in nature, but in an insufficient way, so that another comprehension is necessary. To illustrate the difference between both rhythms, Mondrian distinguishes "the tendency expressed through rhythm established by curved or concentric lines, and the tendency that manifests rhythm through straight lines in rectangular opposition. The first produces undulating rhythm, the second rhythm through cadence" (253). As is easily understandable Mondrian prefers the second one as it corresponds to the privilege he gives to the straight line in his paintings, whereas the first is too "particular" and makes one think of the undulation produced when we throw a stone into a still surface of water, hence too close to nature, as he precisely points out (253).

It is important to stress again that the distinction between particular (natural) and universal (neoplastic) is essential for Mondrian. From this point of view, I cannot follow Tosaki when he claims that "Mondrian gradually introduced a time- and repetition-based 'naturalistic' rhythm into both his painting and into his theory of rhythm. This he later called 'dynamic equilibrium'" (*Mondrian's Philosophy* 4–5). If there is indeed rhythm in nature, this rhythm must be carefully distinguished from rhythm in his neoplastic works. Reintroducing a natural conception of rhythm would be completely contradictory with Mondrian's trajectory. The trouble, maybe, comes from Tosaki's understanding of the relationship between "natural" and "artificial" as being (supposedly) dialectical: "rhythm always exists both in the natural and the artificial, in a dialectic process" (*Mondrian's Philosophy* 70; see also 45). Such interpretation arises from Tosaki's understanding of Mondrian as Hegelian (according to him, besides theosophy, Hegel would be the main philosophical source of Mondrian—Tosaki, *Mondrian's Philosophy* 11–13).

In fact, there is no dialectic but a strong opposition between rhythm in nature and in art, as I have tried to show. In a later text, from 1930, Mondrian explains that the new plastic he promotes, neoplasticism, "*is the clear realization of free rhythm, universal rhythm, which is distorted and concealed in the individual rhythm of form*" (*The New Art* 230, original emphasis). In some rare occasions, however, rhythm is positively related to nature, as in a passage of a writing dedicated to the Italian Futurists' *Bruiteurs* (noise-makers): "In composition, the invariant (the spiritual) is expressed by *straight line and planes of noncolor (white, black, and gray)*, while the variable (the natural) is expressed by *color planes and by rhythm*" (150, original emphasis). Interestingly, it is exactly this sentence, which constitutes an exception in Mondrian's thought, that Kandinsky quotes in his teaching at the Bauhaus (45), because it linked colors and rhythm to nature,[3] a topic in which Kandinsky, unlike Mondrian, was deeply interested.

Even though Mondrian never abandoned the opposition between a negative and a positive idea of rhythm, he nevertheless extended the positive one from art to life in another text:

> Today, it is not only the fine arts that show the rhythm of basic oppositions; the new music and especially American jazz, as well as modern dancing, also tend to express it emphatically. The content of our time is to experience the value

[3] For a comparaison between Kandinsky and Mondrian about music, see Tosaki, *Mondrian's Philosophy* 136–40.

of concrete, practical life; let us try to perfect it by realizing what is essential in human life; *let us try to realize its profound rhythm everywhere in life.* (*The New Art* 252, original emphasis, translation slightly modified[4])

Hence, Mondrian's interest for jazz and its syncopated rhythm (Boccadoro). As is well-known, some of his last paintings are directly inspired by boogie-woogie: *Boogie Woogie*, 1941–2 (B299-319—see Joosten); *Broadway Boogie Woogie*, 1942–3 (B323); *Victory Boogie Woogie* (B324; unfinished). Mondrian saw in jazz part of the "rhythm liberation" he was emphasizing in art, but also in life. According to Mondrian, art is not isolated because there is a movement from art and from life to "true" rhythm: "Just as art comes closer and closer to pure expression of its true rhythm, life also comes closer to the realization of the true rhythm of life" (*The New Art* 255, translation slightly modified). In this particular case, music and jazz in particular helped him to find a free rhythm liberated from the particular rhythm attached to nature and based on repetition and undulation, a free rhythm therefore capable of having a kind of universal validity.

Paul Klee

The artist who most worked on rhythm in the 1920s is indubitably Paul Klee. His thoughts are complex and not easy to summarize in a few paragraphs.[5] The main feature of structural rhythm according to Klee is repetition: "the moment of repetition (*Repetition*) is characteristic of structural rhythms" (*Beiträge* 55). Such a structure is a repetition of pure units added to each other like 1 +2 +1 +2 +1 +2 +1 +2 (43; Figure 11.3). As we saw, Kandinsky held the same. Yet for Klee such rhythms are not organic, as an organism is considered an individual which means that it is not divisible (47). Here comes a difference that is crucial in Klee's thinking, between "dividual" (*dividuel*) and individual; the first is divisible, the second is not. For this reason, structural rhythms correspond to divisible numbers (2, 4, 6, and so on), whereas prime numbers correspond to individuals (3, 5, 7, 11, and so on) (55). Insofar as prime numbers are not

[4] According to the original manuscript in French (Mondrian, "L'art nouveau").

[5] It would be easier to refer here to the English translation of the volume *Das bildernische Denken* (*Notebooks*). However, the edition of this book, due to Jürg Spiller, has been strongly criticized and is not reliable enough (Klee, *Cours* 21). For this reason, I prefer to refer as often as possible to the facsimile of the notes taken by Klee for his classes at the Bauhaus in 1921–2, and reprinted in *Beiträge*; in what follows, the pages given are that of Klee's manuscript in German.

Figure 11.3 Paul Klee, "Rhythmische (Rhythmic)," 1930, oil on canvas, 68.9 × 0.48 cm. Musée d'art modern, Centre Georges Pompidou, Paris. Public domain.

divisible, individuals don't show any repetition (57). However, Klee explains that these prime numbers must be taken qualitatively and not quantitatively (57).

We could logically conclude that there is no rhythm in individuals, as they don't show any repetition. However, things are more complicated as Klee considers the existence of "individual rhythms" (*Beiträge* 58). If it is clear that repetition is the characteristic of dividual or structural rhythms, not of all rhythms; however, the nature of individual rhythm is not as clear.

Klee insists on different kind of possible relationships between dividual and individual and gives the example of "Columbus' fish" (*Beiträge* 56) as "an intimate fusion of individual and structure" (*Notebooks* 297). However, fusion is just one option among many others. Generally speaking the individual belongs to a higher level, whereas the dividual rhythms belong to the lower unit, which the fish illustrates again: "the fish seen as an individual breaks down into head, body, tail, and fins. Seen dividually it breaks down into scales and the structure of the fins" (264). Klee explores so many possibilities of combinations that they can hardly be summarized. There are loose and rigid structures and rhythms (304). In the same way, individuals can be dominant, delicate, or happy (Klee, *Beiträge* 60).

Klee also distinguishes "Rhythm-measure" and "rhythmic weight" (*Beiträge* 50), opposing different measures of weight (light, half-heavy, heavy, very heavy). Furthermore, it is striking that he seeks for rhythm everywhere, that is, above all outside painting. The same as his fellow painters, Klee has a wide concept of rhythm as organic: "We can perceive rhythm with three senses at once. First, we can hear it, secondly see it, thirdly feel it in our muscles. This is what give it such power over our organism" (*Notebooks* 267). He went so far as speaking of "rhythm of matter" (for example, tension between muscle and bone) (Klee, *Beiträge* 80).

Conclusion

I have chosen the five painters examined above because I thought that in the beginning of the abstract movement rhythm had to be important as a way of analyzing nonfigurative images. That's true and so it is. However, there is an additional fact: the period 1910–30 (to which the texts analyzed are limited) belong to what has been considered a "Golden Age" of rhythm, mainly in films (Guido). This means that beyond the particular case of abstract art, a strong interest for rhythm was perceived at the time.

It is also remarkable to realize that the artists under consideration shared a common background. It is out of the scope of this essay to sketch it. However, one name at least must be mentioned, that of Émile Jacques-Dalcroze, a Swiss composer, musician, and educator whose "eurhythmics" had a pervasive influence at the time (Vicovanu 216–28; Pierre). In 1926, he organized the First International Congress on Rhythm at his Jacques-Dalcroze Institute in Geneva. By the way, Kandinsky noted his name down about the relationship between rhythm and danse (*Cours* 150), and Kupka's views are very closed to that developed by Jacques-Dalcroze (Pierre). As far as Mondrian is concerned, he also referred to Jacques-Dalcroze who tried "to create rhythm" starting "from music" (Mondrian, *The New Art* 127). According to the Swiss promotor, the "rhythmic exercise ('gymnastique')" he promoted was based on movement: "every rhythm is movement" (Jacques-Dalcroze, *Méthode Jacques-Dalcroze* VIII), which is also the case for music, as he clearly stresses: "Musical rhythm is like the reflection of bodily movements and we consider it as dependent on the good balance and the general harmony of these movements" (XI). Even though his ideas are more related to music and dance, they obviously had a great impact on painters and stressed the general importance of rhythm.

At the beginning of this chapter I was wondering whether it is possible to see rhythms in paintings. Yet the artists examined all suggest ways of describing and analyzing them. However, they do so insofar as their concept of rhythm goes beyond rhythm limited to the canvas, as it involves movement, breath, and more generally an organic and even cosmic idea of rhythm that might be explained by their first- or second-hand knowledge of Jacques-Dalcroze and his followers, all stressing the importance of bodily (and cosmic) movements to perceive and understand rhythms. The ways our painters conceive of rhythm are very different, so we cannot draw general conclusions. For instance, repetition, symmetry, and proportions are considered as features of rhythm in painting by Kupka, Kandinsky, and Klee, unlike Mondrian. Not surprisingly, however, most of them share a strong interest for the relationship between music and painting: above all Klee, Mondrian, Kandinsky, and Survage. Such interest is easily understandable given the importance of rhythm in music, but this doesn't mean that rhythm in painting would be a mere metaphor; music just helped them to structure rhythm in their own field: the fact that they all felt it necessary to use the idea of rhythm to better understand their own paintings is in itself an important outcome.

Works Cited

Boccadoro, Brenno. "La frénésie du jazz qui vient de s'emparer de notre jeunesse se traite avec du plomb." Le jazz dans le regard de ses ennemis au début du XXe siècle. *Le Rythme. XVIIes entretiens de La Garenne Lemot.* Edited by Jackie Pigeot. Presses Universitaires de Rennes, 2014, pp. 1–35.

Delaunay, Robert. *Rythmes sans fin*, exhib. cat. Centre Georges Pompidou, 2014.

Delaunay, Sonia. *Les couleurs de l'abstraction*, exhib. cat. Musée d'art moderne de la Ville de Paris, 2014.

Enríquez, Flor Marlene, and Paulo Eduardo Lopes. "Ritmo y pintura: notas hacia un acercamiento semiótico." *Acta poética*, vol. 13, nos. 1–2, 1992, pp. 141–67.

Fantini, Bernardino. "Rythmes corporels, rythmes psychologiques, rythmes culturels." *Le Rythme. XVIIes entretiens de La Garenne Lemot*. Edited by Jackie Pigeot. Presses Universitaires de Rennes, 2014, pp. 207–38.

Gaiger, Jason. "Pictorial Experience and the Perception of Rhythm." *The Philosophy of Rhythm: Aesthetics, Music, Poetics*. Edited by Peter Cheyne, Andy Hamilton, and Max Paddison. 2019, pp. 307–30.

Guido, Laurent. *L'Âge du rythme. Cinéma, musicalité et culture du corps dans les théories françaises des années 1910–1930*. L'Âge d'Homme, 2014.

Jacques-Dalcroze, Émile. *Méthode Jacques-Dalcroze. Ire partie. Gymnastique rythmique*, vol. 1. Jobin & Cie, 1906–13.

Joosten, Joop M. *Piet Mondrian. Catalogue raisonné*, vol. 2, *Catalogue Raisonné of the Work 1911–1941*. Harry N. Abrams, 1998.

Kandinsky, Wassily. *Complete Writings on Art*. Edited by Kenneth C. Lindsay and Peter Vergo. Da Capo Press, 1994.

Kandinsky, Wassily. *Cours du Bauhaus: Introduction à l'art moderne*. Denoël/Gonthier, 1975.

Klee, Paul. *Beiträge zur bildnerischen Formlehre: Faksimilierte Ausgabe des Originalmanuskripts von Paul Klees erstem Vortragszyklus am staatlichen Bauhaus Weimar 1921–1922*. Edited by Jürgen Glaesemer.... Schwabe Verlag, 1979.

Klee, Paul. *Cours du Bauhaus. Weimar 1921–1922. Contributions à la théorie de la forme picturale*. Éditions des Musées de Strasbourg and Éditions Hazan, 2004.

Klee, Paul. *The Notebooks of Paul Klee. Volume 1. The Thinking Eye*. Wittenborn Art Books, 2013. Digital Edition.

Kupka, Frantisek. *La Création dans les arts plastiques* (*c.* 1913). Cercle d'art, 1989.

Mondrian, Piet. *The New Art—The New Life. The Collected Writings of Piet Mondrian*. Edited by Harry Holtzman and Martin S. James. Da Capo Press, 1993.

Mondrian, Piet. "L'art nouveau—la vie nouvelle," edited by Jean-Claude Lebensztejn. *Les Cahiers du Musée national d'art moderne*, nos. 114–15, 2010–11, pp. 33–70.

Pierre, Arnauld, "La musique des gestes. Sens du mouvement et images motrices dans les débuts de l'abstraction." *Aux origines de l'abstraction 1800–1914*, exhib. cat. Musée d'Orsay, 2003, pp. 85–101.

Roque, Georges. *Qu'est-ce que l'art abstrait? Une histoire de l'abstraction en peinture (1860–1960)*. Gallimard, 2003.

Survage, Léopold. *Écrits sur la peinture*. L'Archipel, 1992.

Tosaki, Eiichi. "Rhythm as Schema: In Search of the Mature Theory of Rhythm in Visual Art." *Sensorium: Aesthetics, Art, Life*. Edited by Barbara Bolt, Felicity Colman, Graham Jones, and Ashley Woodward. Cambridge Scholars P, 2007, pp. 108–23.

Tosaki, Eiichi. *Mondrian's Philosophy of Visual Rhythm*. Springer, 2017.

Vicovanu, Roxana, "De la 'Grammaire du geste' au 'geste de l'art'. La contribution d'Adolphe Appia et d'Émile Jacques-Dalcroze à une définition du rythme et du mouvement vivants." *Vers la science de l'art. L'Esthétique scientifique en France 1857–1937*. Edited by Jacqueline Lichtenstein, Carole Magné and Arnauld Pierre. Presses de l'Université Paris-Sorbonne, 2013, pp. 211–28.

Visnadi, Débora. "Kandinsky, Malevich e Mondrian: concepção rítmica no surgimento da arte abstrata." *Revista de artes visuais*, vol. 24, no. 40, 2019, pp. 1–18.

Contributors

Nick Admussen, a specialist in contemporary Chinese, earned an MFA at Washington University in St. Louis and a PhD in East Asian Literature from Princeton University. He is an associate professor in the Department of Asian Studies at Cornell University, and published his first monograph, *Recite and Refuse: Contemporary Chinese Prose Poetry*, with Hawaii University Press (2016). He most recently received the 2024 Morgan Chia-Wen Sze and Bobbi Josephine Hernandez Distinguished Teaching Prize.

Chiara Bassetti is Associate Professor at the Department of Sociology and Social Research of the University of Trento. She is Deputy Coordinator of the Doctoral Programme in Sociology and Social Research and Vice-Chair of the homonym MA Programme and an expert in ethnography, ethnomethodology, and conversation analysis. Her chapter focuses empirically on the performative, embodied, aesthetic, and affective aspects of social interaction, and the role of artifacts and technologies, particularly at the workplace and in learning. Empirically, her research ranges from creative practices and performing arts such as dance and music, to complex sociotechnical systems such as emergency centers and airport security, to the home, mundane interaction, and intimate relationships. She has also led projects exploring collaboration in grassroots initiatives for socioeconomic innovation and currently participates to the European Doctoral Network "Reworlding." She often works in interdisciplinary teams. She is Co-Editor in Chief of the journal *Etnografia e Ricerca Qualitativa*.

Yolaine Escande is Research Director at the Centre Nationale de Researches Scientifique (Paris), Deputy Director of the Center for Research on Arts and Language and Professor in L'École des Hautes Ẽtudes de Sciences Sociales, and past member of the Advisory Committee for Works of Art of UNESCO. She specializes in the practice and theory of Chinese graphic arts, calligraphy, and painting. Her numerous publications include *The Landscape between Art and Nature, Sketch, Schema, and Aesthetics, Skiff on the Ocean of Painting,* and *Multiculturality, Materiality and Contemporary Art.*

Anna Gibbs is an associate professor in the School of Communication Arts at the University of Western Sydney and is clinically trained in psychodynamic psychotherapy. Her work combines affect theory, infant research, and feminist theories of the body to analyze mimetic communication in the social field. She recently completed work funded by the Australian Research Council focused on "The Power of the Image: Affect, Audience and Disturbing Imagery."

Nina Kraus is a Professor of Neurobiology and Otolaryngology in the School of Communication and Director and Fellow of the Hugh Knowles Center at Northwestern University. Her revolutionary book *Of Sound Mind: How Are Brain Constructs a Meaningful Sonic World?* was published by MIT Press (2021), and her essays have appeared in virtually every major journal in the field, including *Biology, Journal of Concussion, Journal of Hearing, Auditory Cognitive Neuroscience, Child Neuropsychology, Clinical Neuroscience*, and *Developmental Cognitive Neuroscience fields*. Her twenty years of funded research include support from every major national funder of sponsored research including most recently a Guggenheim.

Keith D. Leonard is an associate professor in and former Chair of the Literature Department and directed the Frederick Douglass Distinguished Scholars program at American University. His much-lauded and widely reviewed monograph *Fettered Genius: The African American Bardic Poet from Slavery to Civil Rights* was published in 2006 by the University of Virginia Press. His more recent essay, "Postmodern Soul: The Innovative Nostalgia of Thomas Sayers Ellis," published in *Contemporary Literature*, received the L.S. Dembo prize for best essay published in the journal (2015).

Caroline Elizabeth Levine (David and Kathleen Ryan Professor) has made distinguished contributions to several facets of nineteenth-century studies, primarily through her well-received books, including *The Serious Pleasures of Suspense: Victorian Realism and Narrative Doubt* (University of Virginia Press, 2003) and *Forms: Whole, Rhythm, Hierarchy, Network* (Princeton University Press, 2015). Her most recent work, *The Activist Humanist: Form and Method in the Climate Crisis* (Princeton University Press, 2023), has turned toward climate and ecological crises. Her scholarly awards include the Perkins Prize for the "best book in narrative studies" (2004) from the Narrative Society, the James Russell

Lowell Prize (2015) and the Dorothy Lee Award for Outstanding Scholarship in the Ecology of Culture, Media Ecology Association (2016).

Mark Lussier. As an emeritus professor of literature and sustainability, an affiliated faculty member in Arts, Media & Engineering and in Religious Studies, and as an administrator, Mark Lussier contributed to every level at Arizona State University and the wider profession. He received special recognition by the Parent Association Professor of the Year and the Zebulon Award for outstanding teacher in the College of Liberal Arts & Sciences. He served on the boards of the North American Society for the Study of Romanticism and the International Conference on Romanticism and hosted their annual conferences. His award-winning monographs *Romantic Dynamics: The Poetics of Physicality* and *Romantic Dharma: The Emergence of Buddhism into Nineteenth-Century Europe* contributed to the transdisciplinary understanding of eighteenth- and nineteenth-century studies, as did his edited collections, including *Engaged Romanticism, Romanticism and Buddhism, Romanticism and Physicality*, and *Reading Blake/Blake Reading.*

Georges Roque served as Director of Research (in the arts) and continues as emeritus faculty member with the Centre National de la Recherche Scientifique (in Paris). His books shifted critical focus to color and rhythm and include *Ceci n'est pas un Magritte: Essai sur Magritte et la publicité* (1983), *Art et science de la couleur* (1997), *Qu'est-ce que l'art abstrait? Une histoire de l'abstraction en peinture (1860–1960)* (2003), and most recently *Art et science de la couleur. Chevreul et les peintres de Delacroix à l'abstraction* (2009).

Richard C. Sha is Professor of Literature, an affiliated professor in Philosophy, and an affiliate of the Center for Neuroscience and Behavior at American University. He is the author of three monographs. The recent *Imagination and Science in Romanticism* received the Jean-Pierre Barricelli prize, whereas *Perverse Romanticism: Aesthetics and Sexuality in Britian, 1750–1850* garnered wide praise across a broad range of disciplinary journals, including in the arts, history, neurology, philosophy, psychology, and the broader sciences. His first book was *The Visual and Verbal Sketch in British Romanticism* (Penn Press, 1997). He is also a distinguished (co-)editor of several collections (e.g., *Romanticism and Consciousness Revisited* and *Romanticism and the Emotions*) and numerous journal specials issues (e.g., *European Romantic Review*, *Wordsworth Circles*). He serves on the boards of *The Wordsworth Circle*, NASSR, and ICR. He is currently at work on *Lashed to the Mast: The Cost of Our Emotion Models*, for which he has received a Distinguished Scholars Fellowship at the Center for Innovation and Creativity at the University of Grenoble for 2026–7.

Sha Xin Wei, Professor at the Schools of Arts, Media + Engineering and Complex Adaptive Systems, directs the Synthesis Atelier for transversal art, philosophy, and technology at Arizona State University. He has been a professor at Concordia University as director of the Topological Media Lab, at the European Graduate School and the New Centre for Research & Practice. Sha's core research concerns processualist approaches to ontogenesis and poiesis. His art and scholarly work ranges from gestural media, movement arts, and real-time media installation through experiential design to critical studies and philosophy of technology. Sha has published in philosophy, media arts and sciences, experimental music and performance, science and technology studies, computer science, and human–computer interaction, including the book *Poiesis and Enchantment in Topological Matter* (MIT Press, 2013). He is an associate editor for *AI and Society* and serves on the governing board of *Leonardo*.

Index

www.ingramcontent.com/pod-product-compliance
Ingram Content Group UK Ltd.
Pitfield, Milton Keynes, MK11 3LW, UK
UKHW020215290926
15470UKWH00001B/210